D1293016

# ROOT CELLARING

**Other Bubel Books:**

Vegetables Money Can't Buy

Working Wood

The Seed-Starter's Handbook

The Adventurous Gardener

## The Simple No-Processing Way to Store Fruits and Vegetables

**by Mike & Nancy Bubel**

**Rodale Press,** Emmaus, Pennsylvania

Book design by Jerry O'Brien
Illustrations by Keith Heberling
Photos by Mike Bubel

Printed in the United States of America on recycled paper containing a high percentage of de-inked fiber.

**Library of Congress Cataloging in Publication Data**

Bubel, Mike.
Root cellaring.

Bibliography: p.
Includes index.
1. Vegetables—Storage. 2. Fruit—Storage. 3. Vegetable gardening. 4. Root cellars. 5. Food—Storage. I. Bubel, Nancy, joint author. II. Title.
TX612.V4B8 641.4'52 79-17233
ISBN 0-87857-277-5 hardcover

10 9 hardcover

Grateful acknowledgment is made to the following publishers for permission to reprint copyrighted material from the following sources:

*Countryside* Magazine: "Energy-Free Food Storage" by Jerry Minnich. Copyright October 1977. Reprinted by permission.
Bookcraft Inc.: *Passport to Survival* by Esther Dickey. Copyright 1969 by Esther Dickey. Reprinted by permission.
Harcourt Brace Jovanovich, Inc.: *Food for Thought* by Robert Farrar Capon. Copyright 1978. Reprinted by permission.
Houghton Mifflin Co.: *Basic Baskets* by Mara Cary. Copyright 1975 by Mara Cary. Reprinted by permission.
Funk and Wagnalls: *Diary of an Early American Boy* by Eric Sloane. Copyright 1962 by Eric Sloane. Reprinted by permission.
Doubleday and Company Inc.: *My Friend the Garden* by Fernand Lequenne. Copyright 1965 by Fernand Lequenne. Reprinted by permission.
Doubleday and Company Inc.: *The Countryman's Year* by David Grayson. Copyright 1932 by David Grayson. Reprinted by permission.
Shambala Publications, Inc., 1123 Spruce Street, Boulder, Colorado 80301: *Tassajara Cooking* by Edward Espe Brown. Copyright 1973 by Edward Espe Brown. Reprinted by special arrangement.
Crabapple Press, Meadville, Pennsylvania: "The Farm Cellar" by Eupha Shanly, page 86, *Poor Joe's Pennsylvania Almanack 1979*. Reprinted by permission.

Grateful acknowledgment is made to William H. Matchett for permission to reprint a portion of his poem "Packing a Photograph from Firenze," from *Water Ouzel and Other Poems*, published by Houghton Mifflin. Copyright 1955 by William H. Matchett. Reprinted by permission.

We'd like to dedicate this book

to the memory of Mike's mother,

Mary Bubel.

This time, like all times, is a very good one if we but know what to do with it.

Ralph Waldo Emerson

# Contents

# Preface

Slowly we are carving a new lifestyle. To some it might seem to be one that is looking backward, for it cherishes the homely, the rude, the unpackaged, the unmechanized, the careful. We do not think of it as a blind shutting out of any visions of the future, but rather, for us, the right way to face the future. The carving is not easy. It is often painful. But in it are the seeds of sanity, of joy.

Mara Cary
*Basic Baskets*

Our interest in root cellars goes way back—to the years when our children were young and we were driving the back roads of central Pennsylvania's lovely Buffalo Valley in search of a farm to buy. On one of our rambles, we discovered a real beauty—a classic stone-faced root cellar dug into a hill. We wished, as only a land-hungry young couple can, that it belonged to us. In a way, that splendid old root cellar *does* belong to us, because Mike took a picture of it—a picture that we still enjoy looking at now that we've found some land of our own and settled in. In another sense, the cellar belongs to us because living with that picture has influenced us. It has said something to us about forethought, preparation, and generous provisioning—and about building a pleasing structure for even the most utilitarian purpose.

We have, in fact, come to see the root cellar as a true expression of folk craft—a thing people make to serve their everyday needs. It is one of the few (mostly) unregulated things you can build in most communities today. The appropriate use of found materials to suit a particular site and family delights us with its varied, sturdy, homey results.

This is the old root cellar in Buffalo Valley, Pennsylvania, that got us started collecting root cellars.

Although we've done the usual sightseeing with our family—Williamsburg, the Smithsonian, the Liberty Bell, Yellowstone Park, Broadway, and the beach—we must confess that what really has most often intrigued us in our travels has been the little backyard gardens, homemade birdhouses, and north-slope root cellars we've seen along the way. When we finally realized that *these* are the things that have meaning for us, we started to take short rambling trips to search them out.

Our previous rather casual game of collecting root cellar ideas intensified when we started to plan a brand-new root cellar for the small house we intended to build here on our farm. In our search for the best root cellar plans, we consulted books and fellow gardeners and met many warm and interesting people along the way. We found so much good information that we decided it should be shared.

Thus this book has grown out of our interest in what people are really doing, on their own, to keep vegetables and fruits from the fall harvest for winter eating. We will tell you what a conventional root cellar looks like and show you how to build one, but we will also show you all kinds of improvised and ingenious systems that people have figured out for themselves, and that work.

There is something about a root cellar, for those who have experienced childhood winters of dependence on such stored bounty, that calls up associations of "home" and "security." More than a few of our older informants grew misty eyed as they told us about root cellars they had hand-dug or once owned (including a built-for-the-ages masterpiece roofed by a century-old brick arch), or to which they had been sent as children to bring up the potatoes for dinner. One young contractor even told us how he left a big boulder exposed in the wall of a basement room of a house he had built because he remembered his grandfather's stone-walled root cellar and hoped that perhaps someone would appreciate that room.

In our own experiments with storing live winter vegetables we've rediscovered another pleasure—akin to the contentment of dressing by the warm fire on a cold morning—and that's the pleasure of delving into one's own store for food to put on the winter table. As we shiver our way back to the warm kitchen with parsnips in our pockets, a handful of potatoes, and a bag of carrots, we feel very good about it all because we've managed to grow it and keep it. We wish each of you the same kind of satisfaction.

Mike and Nancy Bubel
Wellsville, Pa.

# Introduction

Our children . . . should enter adulthood with a basic knowledge of how to store food over winter without the cooperation of a nuclear power plant a hundred miles away. Every animal in the forest is taught this skill; we owe our children no less.

Jerry Minnich
"Energy-Free Food Storage," *Countryside*

Root cellars are as useful today as they ever were. In fact, root cellars in all their forms are as up-to-date as tomorrow, now that costs of food and power used for processing are higher with each passing year. As we see it, the root cellar is right up there with wood heat, bicycles, and backyard gardens as a simple, low-technology way of living well—independently.

The term "root cellar," as we are using it here, includes the whole range of ingenious vegetable-saving techniques from hillside caves to garden trenches. The traditional root cellar is an underground storage space for vegetables and fruits. Where space and lay of the land permit, these cellars are sometimes dug into a hill and then lined with brick, stone, or concrete block. Dirt-floored or insulated basement rooms, somewhat less picturesque but probably more numerous, are also traditional.

Since our purpose in writing this book is to help you to store as much garden produce as possible without processing, we'll also include as root cellaring techniques suggestions for decentralized vegetable storage—in garages, porches, buried boxes, and even right in the garden row, as well as a few ways to keep your family in fresh green vegetables during the winter, even if you don't have a greenhouse.

What would a root cellar do for you? Simply this: Make it possible for you to enjoy fresh endive in December; tender, savory Chinese cabbage in January; juicy apples in February; crisp fresh carrots in March; and sturdy unsprayed potatoes in April—all without boiling a jar, blanching a vegetable, or filling a freezer bag. A root cellar can save you time, money, and supplies. I discovered this the summer we started to build our house. In planning our garden that year, I had to come to terms with the fact that I'd be too busy, as the carpenter's assistant, to do any freezing or canning of garden produce. So I planted vegetables—like tomatoes and corn—that we could eat fresh throughout the summer, and others—parsnips, carrots, and the cabbage clan—that I could harvest in the fall and keep in our cold basement for winter eating. This plan, born of necessity, worked beautifully. Our gas and electric bills were lower because I was not heating two-gallon kettles of water to can things, I was stuffing less into the freezer, and I didn't need to buy new canning jar lids or freezer bags.

We found, too, that root cellaring led us to a whole new system of eating, one based on the age-old seasonal swings. In June we really appreciated the peas because we knew we wouldn't have them in January. In the fall, when frost jewelled the grass and the pig was ready to butcher, we were hungry once again for the hearty, earthy flavor of turnips and rutabagas, beets and carrots, and parsnips. I don't mean to imply that I've quit canning and freezing. I would truly miss my freezer, and our favorite canned goods—tomatoes, pickles, catsup, and peaches—are a must. But I can see now that I was processing more food than necessary, perhaps because of some innate squirrelling instinct that whispered in my ear every August: "Provide, provide." Now I give that impulse a more satisfactory expression by putting by a carefully planned store of winter keepers that make our January meals as special in their own way as those we enjoy in July.

Last evening, for example, I took a basket to our cellar to go "shopping" for the ingredients of the evening meal. Five potatoes, dusty but still firm, filled the bottom of the basket. A fistful of carrots and a single huge beet leaned against the side. Good sturdy root vegetables—just what you'd expect from a root cellar. But there's more. Salad was on the menu too, so I put a long, solid head of Chinese cabbage and a rosy crisp radish into the basket. While in the cellar, shivering a bit out of range of the wood stove, I checked on the witloof chicory sprouts growing in a box of earth by the wall. Looks like they'll be ready for next week's salads. On my way up the stairs I grabbed an onion from the net bag hanging above the stairway.

As I scrubbed the potatoes and chopped the leafy cabbage into the salad bowl, I thought about this direct, earthy, and deeply satisfying connection between our summer efforts in the garden and our winter need for fresh wholesome food. The simple life? I suppose you might say so. It is simply a matter of planning, fertilizing, planting, weeding, watering, weeding, weeding, and weeding, then harvesting and storing away. Snow is predicted for tonight, a thick snow that will drift across the lane and make driving tricky. But there's no frantic dash to the grocery store for stuff to tide us over. We're free to stay home and crack walnuts by the fireplace. Even if the power should go out, so that we wouldn't open the freezer, we could still have a different menu tomorrow, thanks to the root cellar and our home-produced milk and eggs.

Then there is that little corner of our minds that sometimes says "What if?" What if the economy sags to the depths some forecasters predict? What if electricity becomes prohibitively expensive? What if we didn't have our freezer? Could we manage? We feel sure that with some hard work and careful planning and good gardening our root cellar could bring us through.

We've already had a benign test of that need. In last winter's snow, our lane was impassable for two months. We sledded what supplies we needed up the quarter-mile stretch from the road. Thanks to our stored food (in the freezer and canning jars as well as the root cellar) we ate very well indeed, and we made plans to store an even wider variety of good keepers this year.

Storage vegetables needn't be limited to those old standbys: carrots, potatoes, and turnips. With a really well-planned root cellaring program, you can feast on Belgian endive in the dead of winter, fresh tomatoes for Christmas, tender dandelion shoots when the ground outside is ringing hard, nuts and apples, pears and sweet potatoes, even cantalope for your Thanksgiving fruit cup.

Those homely old vegetable carbohydrates are being appreciated in a new light these days. Far from being merely inexpensive menu fillers, complex carbohydrates like those found in vegetables have more value than we've given them credit for. They are an excellent source of fiber. According to studies done by Dr. Robert Bruce and Dr. Benjamin Ershoff, and reported in the January 1978 and May 1975 issues of *Prevention,* eating plenty of fiber helps to decrease levels of certain cancer-causing substances passing through the intestinal tract.

Certain vegetables have, in fact, been found to counter a toxic substance directly. According to studies reported in the May 1977 issue of *Organic Gardening,* the following vegetables help to counteract the toxic cancer-causing effect of nitrosamines that are found in ham, bacon, and many other cured products:

- Cabbage
- Cauliflower
- Radishes
- Turnips
- Pumpkins

If you have a ham in your freezer, then you need some cabbage and turnips in your root cellar! And if you have a well-stocked canning shelf, you need some fresh raw foods to provide the valuable though little-understood enzymes that aren't found in cooked or processed foods.

Some root cellar staples have solidly recognized health-building qualities. Onions and garlic have been shown to be effective in lowering blood cholesterol. Pectin, found in apples, quinces, oranges, grapes, and tomatoes, also reduces cholesterol. And garlic, sometimes called "Russian penicillin" only half in jest, protects against infection, making it a fine food to have on your side during the cold and flu season.

Within the last 100 years, the use of fiberless fat, meat, and sugar has increased to the point where these foods often contribute most of our calories. This is not only an unnatural diet, when considered in the light of mankind's long dependence on vegetable foods, but also an unwholesome one. Vegetables are far more than side dishes. They contribute vitamins and fiber and, when eaten raw, they help to stimulate the production of desirable digestive enzymes and provide valuable chewing exercises for the gums.

You can make profitable use of root cellaring techniques even if your garden is small. If you live in town, you can put by a good hoard of winter vegetables by using some of the "here and there" storage plans described in chapter 12.

Even if you must buy some produce, prices of storage vegetables are usually lowest in the fall. If squash is 25 cents a pound at a roadside stand in October, you can be sure that it will cost more than that in the market in January. Food prices rose at least 10 percent in 1978. With inflation proceeding at that rate, it is not hard to see that growing and keeping your own vegetables will yield a higher return than money you put in the bank.

Your root cellar storage, then, can:

- Help you eat better.
- Save you money.
- Conserve dwindling supplies of energy.
- Give you that priceless feeling of security that comes with being prepared.

It is good to be able to provide for yourself, to be prepared for the winter by your own skill and forethought with your own wholesome home-grown produce. If you like to take charge of your life, choose your food with care and live in a simple, self-reliant way, perhaps this should be your next step toward independence. By spending a little effort, you stand to gain a lot.

# SECTION ONE

## Starting Right With Storage Vegetables

# 1

# Planting Crops for Fall Storage

The countrey-man hath a provident and gainfull familie, not one whose necessities must be alwaies furnished out of the shop, nor their table out of the market. His provision is alwaies out of his own store, and agreeable with the season of the yeare.

Antonia de Guevera,
*The Praise and Happiness of the Countrie-life*
1539

Our first attempts at storing vegetables for the winter were more or less incidental. We'd grown a big long row of carrots and beets and had extras left over in the fall when it was time to clean up the garden for the winter. We brought the surplus root crops into our cool, dirt-floored basement room and kept them there in cartons covered with burlap bags. Gradually we learned how much more our garden could do for us. By spending a little more time planning and planting, we found we could produce as many as 33 different kinds of vegetables for winter storage. With this variety, no one vegetable must carry the burden of being a daily staple, and we don't tire of our stored bounty. Most vegetables that we plant for fall storage may be sown as succession crops, following early peas, lettuce, or beans. This practice makes efficient use of both garden space and soil nutrients. For example, nitrogen left in the soil by the peas and beans promotes leafy growth of fall cabbages and kale. The whole process of dovetailing spring, summer, and fall crops can be intricate but most satisfying to work out. "Let's see now, I planted spring peas on the edge of the garden where soil was easy to dig early. Now that they're finished I can put in kale, which should also be at the edge of the plot so it doesn't get plowed up at fall clean-up time, and which needs a good supply of nitrogen."

As we experimented with growing and keeping different vegetables, we discovered what the experts had known all along: that vegetables to be stored keep best if they're harvested at their peak of maturity — neither underdeveloped nor past their prime. Producing vegetables that are harvest-ready when weather has turned cold enough to provide good storage conditions takes a little planning, but the results are worth every minute spent with pencil, paper, and seed catalogues. Luckily, cool fall weather keeps many vegetables on "hold" so that they don't grow as fast as they would in summer, thus providing a comfortable margin. Carrots that are ready to harvest in September are just fine for winter storage. Even cabbage that heads in September will usually hold — in a cold climate — for winter eating.

Many storage vegetables, in fact, grow best during the cool days of early fall. Lettuce, escarole, and corn salad, which would have bolted to seed in July and August, grow crisp and leafy in September and October. Cabbage, collards, and kale put on exuberant new green growth. Parsnips, salsify, and Brussels sprouts show their excellent true flavors only after frost has nipped them.

Light affects the development of some vegetables too. Cole crops like broccoli produce best during short days. Soybeans are also short-day plants. Decreasing day length triggers their flower formation. Special winter radish varieties produce good roots when days are short, unlike their quick-growing summer-radish cousins which may develop only a thin scraggly root when the sun sets early.

Enthusiasm for digging, planting, and even weeding seems to come naturally in the spring. The impulse that in March often carries us away with its insistence that we nurture some green thing, must sometimes be summoned with some will power in June, July, and August when weather is hot, weeds are persistent, and the whole burgeoning garden is crowded with productive life. Fall frosts seem far away in July when one must weed early in the day to avoid the baking sun, but that is just the time when many good fall-producing crops should be started. As we gardeners like to remind each other, part of each season is spent preparing for the next. After a few years, the wheel of the year takes us with it. Once you've pulled home-grown leeks in November and offered fresh, green, root-cellar salad for Christmas dinner, you find yourself ready to make summer plantings too, even if they must be deliberately scheduled and dutifully carried out, perhaps when you don't really "feel like it." Think of it this way: summer plantings carry the garden forward and keep it productive for the second half of the gardening year, the half you miss out on if impulsive spring plantings aren't succeeded by deliberate summer sowings of durable roots and sturdy leaf crops.

This year, for example, my garden has been an orphan of sorts, for I've been too busy helping with house construction to keep it weeded and consistently replanted. I am not proud of the way it looks this fall. Lamb's quarters five feet tall tower over the carrot row. A carpet of cheese mallow has invaded the cabbages. The ducks have munched on one patch of escarole and the other patch is not blanched — I haven't had a chance to tie up the leaves. Nevertheless, the few hours I devoted to renewing the garden in June and July have kept the rows producing. Under all those weeds, I can still find the following vegetables, this second week in November:

- Carrots
- Escarole
- Head lettuce
- Comfrey
- Chinese chives
- Cabbage
- Broccoli
- Brussels sprouts
- Leeks
- Winter radishes
- Beets
- Parsnips
- Kale
- Parsley
- Jerusalem artichokes

In other years, I have also had rutabagas, kohlrabi, salsify, chard, and turnips.

Sometimes it's hard to find room in the summer garden for your fall crops, and, in my experience, young seedlings sometimes struggle when planted in the row in summer's heat. For these reasons, I've gotten into the habit of starting seedlings of many of my fall crops in flats. I keep the flats on the porch where I can water them and tend them until the seedlings are about two inches high with sturdy stems and several pairs of leaves. Then I transplant them into spaces in the garden — often into a just-cleared row where I've pulled onions, and sometimes into spots where an early planting has finished or failed. This gives me an additional two weeks growing time for early crops in the garden and ensures a good start for the seedlings.

The usual transplanting precautions are especially important in summer heat. Whenever possible, I move seedlings into the garden row on a cloudy day. If rain is expected, I find planting just before a shower preferable to planting after rain. One is less likely to compact soft wet soil by walking over it, and the young plant is naturally watered into place. (I always pour a cupful of water into the planting hole too, of course.)

New transplants need some protection from hot sun for their first three to seven days. I often use berry baskets, which cast a light grid of shade but let sun shine through too. A leafy branch also works well, either placed lightly over the seedlings or stuck in the ground to cast a shadow. Summer transplants usually need to be watered several times in addition to the watering you gave them when setting them out. Mulch them as soon as you can, too, to hold moisture in the soil and control weeds. Once I start canning in August, I find that weeds often get ahead of me, so summer mulch is especially good crop insurance.

A fall garden in a small backyard (about September 10). The healthy seedlings illustrate that you don't need much space, just good planning!

When sowing seeds directly in the row, I usually water the seeds in the open furrow and *then* draw a ¼- to ½-inch layer of fine soil over them. This helps to prevent crusting of the soil. Plants that have delicate foliage, like carrots, often appreciate the additional protection of a thin layer of dried grass clippings or fine, light hay scattered over the row.

Summer planting for fall harvests takes a bit of gumption and persistence. But once started, it becomes a habit. The sight of a row of ruffly, blue-green kale plants flanked by just-heading cauliflower, tender crisp fall head lettuce, and wrist-thick leeks, all lightly silvered with dew on a snappy September morning, will do much to confirm the habit and your own respect for yourself as a gardener and provider. I am busy, absent-minded, and sometimes get behind in my weeding. If I can raise this kind of fall abundance, you can too.

# 2

# Good Keepers

. . . Through the improbable winter to the impossible spring.

William H. Matchett
"Packing a Photograph from Firenze"

Here's a checklist to consult at the beginning of the gardening season — which is, for most of us, in January and February when we pull a rocker up to the warm woodstove or glowing fireplace with our lap full of catalogues, ready to fill in the rows in our mind's eye garden — the one that is always perfect. We have a brand-new chance, now, to remedy the mistakes and shortcomings of the last gardening season. A look at the root cellar will tell us whether our order of good keepers should be increased. Old-time gardeners referred to the late winter and early spring weeks as "the hungry gap" — when stored vegetables ran low, fall-butchered meat was used up, the cow was dry, and the hens hadn't resumed laying. If you're left with only a handful of root vegetables at midwinter seed-ordering time, now's your chance to provide for a bigger and longer-lasting winter vegetable harvest next fall. We've kept this list of good keepers separate so that you'll find it easy to refer to when planning your garden. Most of the vegetables discussed in this book will keep well regardless of variety, but for really outstanding storage life, you might want to try some of these especially reliable varieties. Unless specific sources are noted, the variety is widely available.

A fall harvest basket filled with beets, radishes, and celeriac.

**Beets**

Long Season — our favorite keeper beet; huge and rough-looking, but very tender.
Lutz Green Leaf
Hybrid Red Cross (Farmer)
Detroit Dark Red
Perfected Detroit (Gurney)

**Broccoli**

Waltham 29 — a good fall broccoli.
Green Comet (hybrid) — good for a second planting.

**Brussels Sprouts**

Jade Cross
Long Island Improved
Green Pearl (Stokes)
Roodnerf Rido (Thompson and Morgan) — small sprouts but very hardy.

**Burdock**

Takinogawa Long (Johnny's Selected Seeds)

**Cabbage**

Penn State Ballhead
Danish Ballhead
Jupiter (Thompson and Morgan) — very cold-hardy, will stand long in garden in cold weather.
Premium Flat Dutch
Evergreen Ballhead (Stokes)
Storage Red (Stokes)
Mammoth Red Rock (Hart)
Brunswick (Johnny's Selected Seeds)
Amager Green Original (Johnny's Selected Seeds)
April Green (Stokes)

**Cabbage, Chinese** — Wong Bok types keep well.

Wintertime (Stokes)
Summertime (Stokes)
China King
Matsushima (Johnny's Selected Seeds)

**Carrots**

Chantenay
Danvers

Scarlet Keeper (Johnny's Selected Seeds)
Nantes Forto (Johnny's Selected Seeds)
Kinko Chantenay (Johnny's Selected Seeds)
Gold Pak
Rouge de Tilques (Johnny's Selected Seeds)
Spartan Bonus Hybrid

**Cauliflower**

Newton Seale (Thompson and Morgan) — has extra frost resistance and can remain longer in garden row.

**Celeriac**

Alabaster (Burpee)
Marble Ball (Harris)
Large Smooth Prague

**Celery**

Giant Pascal
Utah
Fordhook

**Colbaga (Farmer)** — according to the catalogue, colbaga "combines the flavors of Chinese cabbage, cabbage, and rutabaga."

**Collards** — Vates is good because it is shorter, but all collards are very hardy.

**Eggplant** — not a long keeper, but try these late varieties.

Jersey King
Burpee Hybrid
Imperial Black Beauty

### Endive

Salad King (Stokes, Field, DeGiorgi) — frost resistant.

### Escarole

Batavian Full Heart
Florida Deep Heart (Herbst)

### Kale

Dwarf Siberian
Dwarf Blue Curled Vates
Green Curled Scotch
Harvester LD (Johnny's Selected Seeds)
Konserva LD (Johnny's Selected Seeds

### Kohlrabi

Grand Duke
White Vienna

### Leeks

Conqueror (Harris)
Elephant
American Flag
Musselburgh
Siegfried (Johnny's Selected Seeds)

### Onions

Yellow Globe Danvers
White Portugal (Burpee)
Buccaneer (Harris)

Burpee Yellow Globe Hybrid
Elite
Chieftain (Stokes) — a Spanish onion.
Ebenezer
Autumn Splendor (Olds)
Southport Yellow Globe (Farmer)
Downing Yellow Globe (Field and Gurney)
Spartan Sleeper (Thompson and Morgan) — said to keep well even in a warm kitchen.

**Parsnips**

Hollow Crown
Offenham
Harris' Model
All-America

**Potatoes, Sweet**

Centennial
Porto Rico
Allgold — our favorite — keeps until spring and high in vitamin A.

**Potatoes, White**

Norgold Russet
Superior (Jung)
Katahdin
Kennebec
Burbank — best grown on light soil.
Snowchip
Red la Soda — medium late.
Sebago — late.

Note: Cascade and Norchip are *poor* keepers.

**Radishes, Winter**

China Rose
Round Black Spanish
Chinese White or Celestial
Miyashige (Johnny's Selected Seeds)
Shogoin (Redwood)
Ohkura (Redwood)
Sakurajima (Redwood and Shumway)

**Rutabaga**

Altasweet (Stokes) — good flavor.
Laurentian
Macomber
Purple-Top

**Salsify**

Sandwich Island, the standard, is offered by all seed sellers who sell salsify.

**Squash**

Blue Hubbard
Acorn — fair keeper.
Butternut — especially Waltham and Hercules strains.
Buttercup — a delicious squash.
Vegetable Spaghetti — not advertised as an especially good keeper, but keeps well for us, till midwinter.
Gold Nugget
Sweet Meat (Harris)
Delicata or Sweet Potato (Stokes)
Guatemalan Blue Squash (Nichols) — an ancient variety
Sakata (Nichols)
Improved Green Hubbard (Farmer)
Melon Squash (Thompson and Morgan, Jung, Gurney, Shumway) — see chapter 9, "Fruits."

### Tomatoes

Burpee is working on a new tomato for storage, called Long Keeper. It was given as a premium for orders in the 1979 gardening catalogue. Perhaps it will soon be offered for sale.

### Turnips

Purple Top White Globe
Just Right (Harris)
Des Vertus Marteau (Epicure) — delicious flavor.

# 3

# Growers Keepers

## How to Raise Top-Quality Storage Vegetables

Gardening *is* soil management. . . . soil is the most complex substance with which we may ever have to deal.

R. Milton Carleton
*The Small Garden Book*

When you've progressed from storing incidental fall surplus to making purposeful plantings for good eating all winter long, you'll find that planting times and growing conditions can make a difference in the amount and quality of your fall harvest.

With the exception of tomatoes, squash, sweet potatoes, and melons, most storage vegetables thrive in cool weather. Insect problems are usually much less serious after the first frost, so the combination of cool weather and reduced competition can help you to raise some fine fall crops. The time of planting will have a bearing on the quality of the storage vegetables, too. You want to put away vegetables that are mature, ready to serve, but not overage and tough. The information in this chapter should help you to time and nurture your plantings so that the vegetables you count on will be at just the right stage of development when you want to harvest them. If you're an experienced gardener, you know that frost improves the flavor of certain vegetables. Parsnips and salsify, for example, are sweeter after frost because some of their root starch is turned to sugars. Kale, Brussels sprouts, collards, and Chinese cabbage have a much milder and more pleasing flavor after frost.

What is not so widely acknowledged is that the soil in which your plants grow can influence the keeping quality of the vegetables you harvest. According to studies reported in E. P. Shirakov's *Practical Course in Storage and Processing of Fruits and Vegetables,* abundant potash in the soil promotes long storage life of fruits and vegetables grown on that soil. Good sources of potassium include manure (especially sheep, horse, and pig manure), compost, green manures, and rock powders like rock potash and greensand. Seaweed, wood ashes, corncobs, comfrey, peanut hulls, and citrus peels also contribute potash to the soil. What's more, as Jeff Cox wrote in his article "Potash: The Plant-Growth Catalyst," in *Organic Gardening,* potassium can be, in effect, stored and released by organic matter in the soil, which attracts the potassium and keeps it until it's needed by the plants. Good gardening practices that add organic matter to the soil include mulching, applying manure and compost, and turning under special plantings of green manure crops like soybeans, clover, buckwheat, annual rye, or alfalfa.

Organic matter worked into the soil also helps to reduce potato scab and other plant diseases by promoting the growth of beneficial bacteria and fungi which help to control the undesirable ones. Sound, unblemished, disease-free vegetables keep best. Studies have shown that well-fertilized kale, cabbage, and collards contain more vitamin C than those grown on lean soil. For root vegetables, though, you want to avoid heavy doses of high nitrogen fertilizer, because excess nitrogen causes watery growth which is low in carbohydrates and which spoils sooner in storage.

Fruit quality and storage life are also adversely affected by high levels of nitrogen in the soil. Studies indicate that the abundance of nitrogen increases metabolic rates, thus aging the fruit faster. According to studies reported in *Symposium: Postharvest Biology and Handling of Fruits and Vegetables,* fruit well supplied with calcium is less likely to suffer the undesirable effects of excess nitrogen. A sign of calcium deficiency in fruit trees is the development of dead spots at the edges and tips of young leaves, especially on the newer twigs. Good natural sources of calcium include limestone, bone meal, and wood ashes. Pruning and other procedures that stimulate new growth on the tree also affect the balance of calcium in the tree's tissues. If there are a great many new shoots on a tree, they can sometimes use up some of the calcium needed by the developing fruit. That's not to say, of course, that you shouldn't prune, but just a reminder that everything we do to or for our plants has its effect.

Potatoes grown in sandy or sandy-loam soils tend to keep better than those raised on heavy clay. While there's not much you can do to change the basic nature of your soil, you can improve the texture of heavy soil by adding lots of organic matter.

Here are some tips to help you to raise the most and the best of your favorite storage vegetables:

**Beets** Plant seeds for the fall crop in June and July, except for the excellent beet "Long Season," a huge, coarse-looking but tender and fine-fleshed good keeper that should be planted in April or May. Firm the soil well over the seeds when planting beets. Thin the plants to stand four inches apart for best development of the roots. (Beets always need thinning because each of the seeds you plant is actually a corky conglomerate of several individual seeds.) Where beets do poorly, suspect toxic chemicals in the soil, acid soil, nearby walnut trees, or lack of boron.

**Brussels Sprouts** Plant seeds in a nursery row or in flats in May, or no later than early June, and transplant seedlings to their permanent spot when three to four inches high. Drying of roots at transplanting time can make the sprouts that develop later loose and leafy rather than solid, so be sure to water the roots into the hole and press covering soil down firmly. Otherwise the plant is undemanding. Regular good garden soil suits it fine; too much fertilizer promotes leafy top growth at the expense of the sprouts. Some gardeners nip out the central rosette of leaves on the top of the plant in September to send energy into the sprouts budding along the stem. When lower sprouts form, snap off all the lower leaves to give them room to develop. Flavor is best after frost.

**Cabbage** Start fall cabbage plants in May or early June. The plants have shallow roots and need a steady supply of water when young, but little care thereafter until harvest. Fertile soil and occasional doses of manure tea promote leafy growth and, of course, those solid heads are all leaves. Excessive rain can cause splitting when heads are mature. You can often prevent splitting by twisting or gently pulling the stalk to break some of the feeder roots. If cabbage does split, use it promptly. It won't keep.

**Cabbage, Chinese** With the exception of several newly developed spring varieties, Chinese cabbage is a fall plant. Heat and long days make it go to seed. Start your fall supply of this delicate, delicious celery cabbage in July and plant the seeds right in the row where they are to grow because they do not take kindly to transplanting. (I *have* dug up and replanted some seedlings, just to see whether I could; they survived, but I wouldn't count on it for a whole crop. Transplanting also often causes these vegetables to go to seed early.) Thin the cabbages to a 12-to-18-inch spacing and give them rich soil, or several helpings of manure tea in ordinary soil. Then watch as short days and snappy nights bring on the crisp, leafy, cylindrical heads for which this loose-leafed cabbage is enjoyed.

**Cardoon** Plant cardoon seeds indoors in March and treat them like tomato seedlings, setting out plants after the last frost. Set the seedlings three feet apart in trenches and gradually fill in the trenches as the plants develop. This blanches the cardoon heart naturally. In September, blanch the stalks by banking soil around the plants. Wrap or tie the tops of the stalks together to keep dirt out of the center of the plant. Don't omit the blanching as cardoon is very bitter without it.

**Carrots** Sow seed for fall carrots in June or no later than July. Deeply dug soil is good. Thin the roots to two inches apart and keep the row well weeded. A wide row shades out many weeds, uses garden space efficiently, and is easy to harvest. Thick-cored carrots often keep better than those with thin cores. Carrots grown in heavy soil tend to be more fibrous. Early spring-planted carrots often become bitter by late fall if left in the ground. A second planting is best for winter keeping.

**Cauliflower** Start fall seedlings in May and keep them growing steadily. Give them plenty of shade and water at transplanting time. Well-limed soil is best. Cauliflower is — well, not neurotic, but *sensitive.* Any check in growth is likely to retard the plant to the point where it will produce only a button head or none at all. Self-Blanche is excellent for fall eating and, as the name implies, needs no blanching. Purple cauliflower doesn't need blanching either. Other varieties must be blanched. Tie the leaves over the head while the head is still small — about three inches across.

**Celeriac** This agreeable root should be more widely grown. Start plants for storage in midspring and transplant the seedlings into the garden when two to three inches high. They don't mind heavy soil but appreciate a good supply of lime. The root is solid and fist-size, with many gnarled and tangled secondary roots at the bottom. Leaves are like those of celery but taste bitter. For large roots, plant in good rich soil and provide plenty of moisture.

**Celery** Sow the fall crop in April or May. Seed may take three weeks to germinate; keep the seedbed moist. Celery likes rich, well-limed soil and a steady supply of moisture. Its root system is shallow and small. Keep the plants well watered and mulched all season long. Green celery has more vitamins than the pale blanched stalks but if you want to blanch celery, bank the plants with soil or tie newspaper jackets around them for three to four weeks in the fall.

**Collards** Treat like cabbage. The flavor is best after frost.

**Endive and Escarole** Sow seeds for fall escarole and the closely related endive in early July in flats or directly in the row. Thin seedlings to stand a foot apart. These frilly leafed greens do well in ordinary garden soil but especially appreciate a good supply of lime and an occasional dose of manure tea. If you want to blanch the heads, tie the leaves together *when they're dry* (they may rot if wet). Blanching takes about three weeks in the fall.

**Horseradish** Plant the root cutting with the thicker end up in spring or fall. Soil type is not critical; horseradish is adaptable.

**Jerusalem Artichoke** We prize these crisp roots for slicing into salads after summer cucumbers are finished. It's hard to stop Jerusalem artichokes. They flourish even on poor soil and spread enthusiastically. They may be planted in a hedgerow but will be easier to dig in softer cultivated soil. Give them a separate bed, though, or they'll take over your garden! Plant the seed tubers four inches deep and a foot or two apart, depending on how many you have. (Tubers may be cut into several pieces as long as each piece has an "eye" or growing bud.)

**Kale** If you plow or till your garden in fall, plant kale at the edge so it won't be harmed. The thickly curled leaves make an attractive border. Start seedlings in May and June — even July is not too late. You can grow lettuce or other quick-maturing crops between the kale plants in summer. Kale does well in most kinds of soil but is truly excellent in rich ground. We give our row several helpings of manure tea in late summer.

**Kohlrabi** Plant in July for crisp fall bulbs, either in seedling flats (transplanting later) or right in the row. Space the plants four to five inches apart. When thinning, cut the seedlings rather than pulling them to avoid damaging the roots of neighboring plants. Kohlrabi roots tend to tangle.

**Leeks** Here's one fall vegetable you start early — while the dug-up leeks from last fall are still green in your cellar. Start seeds in flats, snipping the lanky growth of the thin spears back to a uniform height of two to three inches, or sow seeds in the garden row in April. Leeks occupy the row for most of the season but their winter hardiness makes them worth the space. Our leeks haven't been bothered by insects or disease and take little attention during the growing season. Deeply dug, rich soil raises wonderful leeks. Master gardener Milton Carleton tells a wonderful tale, in his book, *Vegetables for Today's Gardens,* about the Scotch gardener whose secret of success with leeks was that he buried the Laird's trousers beneath the row.

**Lettuce, Head** Sow the seed in July in flats or the garden row. Sometimes lettuce seed is reluctant to germinate in summer heat, but you can usually persuade it to sprout by chilling the seed in the refrigerator for several days before planting and exposing the planted seed to light. Mulch young plants if August is dry and give them a helping of manure tea when they begin to head up in September. Lettuce has a short storage life, but it lasts well into the fall in the garden. By protecting it from frost we regularly have head lettuce into November.

**Onions** You can grow these from seed or sets (tiny onion bulbs), or you can buy plants by mail from seed companies. Seed-grown onions are especially good for storage. Plant the seed rather thickly in flats in February. Clip the tops to a height of two to three inches. Set out plants four inches apart in the row in April. Sets may be poked into the ground as soon as the soil can be worked. Keep onions well weeded. If they don't get enough water, they'll taste strong. Long summer days promote growth of the bulb, so get those onions planted early.

**Parsley, Hamburg-Rooted** Hamburg-rooted parsley is excellent in soups and stews. This dependable if unglamorous vegetable may be planted right in the row as early as the ground can be worked. Germination may take a month; presoaking the seed can chop a week off this time. Mark the row with a few radish seeds. Thin roots to two inches apart; they'll grow long and slim like carrots.

**Parsnips** Slow to develop but very long lasting, these nutritious roots are planted in April (May in the far north) and harvested in mid or late fall after frost has brought out their sweet flavor. Don't bother to eat parsnips in the summer; their flavor isn't half as good then and you have other food in the garden then that *is* in its prime. Be sure to use fresh seed; parsnip seed over a year old can't be counted on to sprout. I like to plant parsnips in a wide row as I do carrots.

**Peppers** When you find a pepper that does well in your area, stick with it. We're partial to Bell Boy Hybrid. Calwonder does well in warm areas but not in the north. Vinedale, Staddon's Select and Earliest Red Sweet are good for short seasons. Plant seeds indoors in February or March and set seedlings in the garden when frost danger is past. Dry weather can interfere with pollination. Try misting the blossoms. Peppers aren't demanding plants; they tolerate acid soil better than many other vegetables, and they don't need rich soil either. Magnesium does aid in their growth, though, so we use magnesium-containing dolomite limestone to improve our garden soil. Hot peppers, which are closer to the wild form of these South American perennials, accept even heavy clay soil which usually contains too little air for optimum growth of most other vegetables.

**Potatoes, Sweet** Hill up a ridge of soil so that these vitamin-rich roots will have someplace to go when they develop. In a good year, you'll get some huge ones. Sandy soil that warms fast is traditional for sweets, but don't fret if your soil is heavier, more on the clay side. We have excellent crops here on heavyish soil. Sweets are content on relatively poor soil; they don't need high-nitrogen fertilizers. Start plants by rooting last year's sweet potatoes in water in a warm place, or buy plants by mail. Set them a foot apart in that ridge you hoed up along the row. Here in south-central Pennsylvania, we plant our sweet potatoes on the last day in May. If you mulch between the rows, two weedings should carry you through the season.

**Potatoes, White** The nicest potatoes we've ever harvested were grown in a bed of mulch. They were large, clean, and easy to "dig." The catch is that you must use enough mulch — at least 10 to 12 inches — to keep the tubers well covered, or light shining through will turn them green.

Green tubers contain the toxic alkaloid solanine. New potatoes form *above* the seed pieces you plant, not below as you might expect, so it's important to cover the plants well with mulch or soil hilled up from the row middles.

Well-drained, humus-rich soil on the acid side raises good potatoes. Buy certified disease-free seed potatoes or plant your own healthy spuds, but beware of planting shelf-run potatoes from the grocery store, which could carry disease. Although they may be planted as early in spring as the soil can be worked, many gardeners make a second late-spring planting for fall use. Space the seed potatoes or pieces about 12 to 15 inches apart in furrows 4 to 5 inches deep. Sebago and Katahdin are good late varieties. High-nitrogen fertilizer can retard tuber growth if applied at blossoming time while tubers are forming. Cool nights promote storage of starch, making for a good, mealy, long-keeping potato. Rotate the patch to prevent disease build-up.

**Pumpkins** Don't hurry to plant them. What you want is a well-developed storage crop, not a quick harvest. Planting at the end of May allows plenty of time in most areas. Pumpkins like warm weather and rich soil.

**Radishes, Winter** Plant short rows of these pungent appetizers as space becomes available in July and August. They develop large leafy tops as turnips do, so they need somewhat more row space than quick summer radishes.

**Rutabagas** Sow seeds in June or July and thin the seedlings to six inches apart. Early spring plantings seldom do well in summer heat; the plant develops a long, tough neck and a runty root. No nitrogen-rich fertilizers for this lean-soil crop, but well-limed soil is good. Rutabagas grown in boron-deficient soil may have soft brown interior rot. A good band of compost applied to the planting furrow for the next crop should help to remedy this problem. When there's too much raw manure or plant material in the soil, the roots may be infested with wireworms.

**Salsify** The vegetable oyster, as it is often called, should be planted in May to develop its long, slender roots in time for fall digging. Loosely worked soil promotes good root development. Otherwise salsify does well on ordinary garden soil. Flavor is best after frost.

**Scorzonera** Sometimes called "black oyster plant," scorzonera resembles salsify but is not closely related. It has black skin and white inner flesh. The long thin roots do well in deeply dug well-drained soil. It takes three to four months for the roots to grow to usable size. Plant the seeds in May and thin the plants to stand about three inches apart. Deep cultivation close to the developing plants may cause forked roots. Treat like salsify.

**Squash** Treat these like pumpkins, and don't rush to plant them. You want them to mature when storage conditions are good. Late May is a good time to plant. Toss a shovelful of fine garden soil over several vines as they develop so that parts of the plant will continue to live on the new roots that form, even if the main stem is attacked by a squash borer. We put finished compost in the hill when planting squash and pumpkins.

**Swiss Chard** Swiss chard is one of the most goof-proof of all vegetables. It flourishes in a wide variety of soils and does better in acid soil than the beet. It has an exceptionally deep, wide-ranging root system. Plant the seeds about a month before the last frost. After picking from the plant all spring and summer, you can cut down all the straggly old leaves in August and look for a tender new fall harvest in a month, all from the same plants.

**Tomatoes** Late-planted tomatoes are best for storage. You can make a special late planting and nurse along several volunteer plants or root cuttings taken from mature plants in midsummer. A dose of manure tea or other natural plant food will give the plants a boost when fruiting begins. About two or three weeks before frost is due, pinch out blossoms and tiny fruits and nip off the leafy tops of the main branches to spur concentration of nourishment in the larger developing fruits.

**Turnips** Spring-planted turnips shouldn't be counted on for fall storage; they'll be woody and strong flavored. Make a special planting in July or August on good humus-rich soil. You can plant in rows or sow a solid bed, using the thinnings for soup. Turnips are sometimes mishandled because they are easy to grow. Try harvesting them no larger than three inches in diameter.

# The Swing of the Seasons

## How One Family's Good Garden Eating Varies from Month to Month Around the Year

### JANUARY

Fresh Vegetables:

- Kale
- Witloof chicory (Belgian endive)
- Leaf lettuce grown under grow-lights (soon in solar greenhouse)

Frozen Vegetables:

- Peas
- Beans
- Corn
- Broccoli
- Cauliflower
- Greens

Canned Vegetables:

- Tomatoes
- Sauerkraut
- Pickled beets
- Pickled cucumbers
- Pickled beans
- Assorted other pickles and relishes

Root Cellar Vegetables:

- Onions
- Chinese cabbage
- Beets
- Carrots
- Potatoes
- Squash
- Garlic
- Sweet potatoes
- Leeks
- Parsnips
- Salsify
- Rutabagas
- Radishes
- Turnips
- Cabbage
- Celeriac

## FEBRUARY

Same as January, plus Jerusalem artichokes if there is a thaw, rhubarb if sprouted in cellar.

## MARCH

Same as January and February, plus some wild greens and minus the winter radishes, leeks, and rutabagas which are usually used up by now.

## APRIL

Fresh Vegetables:

- Asparagus
- Parsley
- Swiss chard (new growth from last year's plants)
- Wild dandelion, dock, winter cress

Frozen and Canned Vegetables: as above

Root Cellar Vegetables:

Potatoes
Onions, unless used up
Carrots
Beets, sometimes
Leeks, in the garden from last year
Parsnips, wintered over in the garden
Jerusalem artichokes
Sweet potatoes
Garlic
Salsify, in the garden from last fall

**MAY**

Fresh Vegetables:

Asparagus
Sugar peas
Lettuce (leaf)
Spinach
Wild greens
Comfrey
Turnip greens

Frozen and Canned Vegetables: as above. Use up the last frozen vegetables from previous year and clean out freezer.

**JUNE**

Fresh Vegetables:

Asparagus
Beans
Peppers
Lettuce
Peas
Zucchini
New Zealand spinach
Onions
Broccoli

Kohlrabi
Cabbage

Frozen Vegetables: not used in summer

Canned Vegetables: seldom used, except for pickles

Root Cellar: emptied and cleaned out in June

## JULY

Fresh Vegetables: All vegetables available in June, plus corn, tomatoes, cucumbers, carrots, and beets.

Root Cellar: not needed

## AUGUST

Fresh Vegetables: All vegetables available in June and July except asparagus and peas.

## SEPTEMBER

Fresh Vegetables and Fruits:

Tomatoes
Peppers
Corn
Lettuce
Eggplant
Broccoli
Cabbage
Spinach
Zucchini
Cucumbers
Potatoes
Apples
Pears
Escarole
Chinese cabbage
Carrots
Beets

Root Cellar Vegetables and Fruits:

Onions
Garlic
Potatoes
Apples

**OCTOBER**

Fresh Vegetables: All of those available in September except corn. Also:

Winter squash
Parsnips
Salsify
Brussels sprouts
Kale
Collards
Turnips
Leeks
Rutabagas
Winter radishes
Sweet potatoes
Cauliflower
Beans, sometimes

Frozen and Canned Vegetables: as in January

Root Cellar Vegetables and Fruits:

Potatoes
Onions
Apples
Garlic

**NOVEMBER**

Fresh Vegetables in Garden (some protected):

Parsnips
Salsify
Brussels sprouts
Kale

Collards
Turnips
Leeks
Head lettuce
Spinach
Winter radishes
Rutabagas
Broccoli, sometimes
Cauliflower, sometimes
Escarole
Chinese cabbage
Regular cabbage
Carrots

Root Cellar Vegetables:

Beets
Peppers
White potatoes
Sweet potatoes
Onions
Garlic
Cauliflower
Broccoli
Ripening tomatoes

## DECEMBER

Fresh Vegetables in Garden:

Leeks
Kale
Chinese cabbage (early in month)
Parsnips
Carrots
Salsify
Brussels sprouts

Frozen and Canned Vegetables: as in January

Root Cellar Vegetables:

Sweet potatoes
White potatoes
Green and ripe tomatoes
Chinese cabbage (later in month)
Cabbage
Onions
Carrots
Garlic
Winter squash
Escarole
Turnips
Winter radishes
Rutabagas
Parsnips
Jerusalem artichokes
Beets

# How Much Do You Need to Put By?

If you're working with simple garden surplus, naturally you'll store all your extras. But if you're planning ahead for a truly self-sufficient winter, you might find the following guide helpful. Family tastes, ages of children, and work habits of adults—at home or away from home—are personal variables that you'll also need to take into account. If you enjoy a wide variety of storage vegetables, you'll need less of each than if you eat only the basics—potatoes, carrots, beets, and onions. For a family of four, you'll need approximately:

- Beets: 1-2 bushels.
- Carrots: 2-3 bushels.
- Cabbage: about 30 heads.
- Brussels sprouts: about 10-15 plants in garden.
- Chinese cabbage: 20-30 heads.
- Celery: 10-20 stalks.
- Turnips or rutabagas: about 1 bushel of each.
- Potatoes: 6-14 bushels.
- Sweet potatoes: 2 bushels.
- Endive: 10-20 plants for storage, more in the row for late fall.
- Squash and pumpkins: 30-40.
- Onions: 1-2 bushels.
- Parsnips: 1-2 bushels.
- Salsify: ½-1 bushel.
- Leeks: 15-40 plants.
- Celeriac: ½-1 bushel.
- Kale: 50-100-foot row.
- Winter radishes: ½-1 bushel.
- Kohlrabi: ½-1 bushel.
- Garlic: as desired. A 25-foot-long wide row planted 4 cloves across should be plenty. Yield would be approximately ⅓ peck.

# SECTION TWO

## Bringing in the Harvest

# 4

# How to Harvest and Prepare Vegetables for Storage

November 25: The rain was succeeded by a blustering storm out of the north. I went out with my coat buttoned to the chin and lifted the last of the celery and brought it into the root cellar. Almost always, terrified by the blusterings of winter, I have been too early. The celery does better in the ground until the last moment before winter comes roaring down. Of course it is difficult to determine the *last moment.* I thought it was today. (Postscript written December 10: I was quite right.)

David Grayson
*The Countryman's Year*

When heavy frost sparkles and crunches in the grass each morning, we know it's time to begin to think about bringing in the root vegetables. The soft fruits — tomatoes, peppers, and such — have already been picked when the first light frosts struck, and stored away for short-term keeping. Hard-shelled fruits like squash and pumpkins were left in the field to cure after they matured and are now sprawled on the porch, waiting to come in where it's warmer. The roots, bless them, wait patiently underground — and even continue to grow during the weeks of fine weather that often follow the first frost.

It's a good thing that the fall harvest *can* be done in stages, for when at last it is time to bring in the roots, there's a lot to be done — digging, trimming, rounding up containers, gathering sawdust, sand, and leaves, lugging baskets inside, and packing the vegetables away. If you grow a lot of root vegetables as we do, you can count on spending several weekends getting them out of the ground and into storage.

*Timing* of the harvest is important. Although soft fruits should be gathered before frost can nip them, and hard-shelled fruits should be protected from heavy frost, root vegetables may be left in the ground even until black frost — that unmistakable hard freeze that kills off even the last green weeds and blackens the beet tops. It is, in fact, a good idea to leave the root vegetables in the ground as long as you possibly can, so that when you do harvest them the temperature in your storage area will no longer be affected by Indian summer quirks and will stay low enough for good keeping.

Digging carrots that were stored in the garden row.

Experience (translate: doing it wrong the first time) has shown us that there are two things to watch out for here. One is the tendency of some root crops, especially beets, rutabagas, and sometimes turnips, to shoulder their way above ground. When exposed like this, they should be mulched to prevent damage from severe frost. Second, it's awfully easy to get busy with other garden clean-up chores and postpone the harvest *too* long — until the ground is frozen and hard or impossible to dig. Those Indian summer days beguile us into thinking they'll last forever. Watch the weather *and* the calendar. Try to wait out that last lingering spell of mild weather, but be realistic, too. Once November has gotten a foothold, winter's on its way. In the northern parts of New England, the Midwest, and the Great Plains, make that October. In the upper South, though, you can safely wait until December.

Cold weather at digging time prolongs storage life in another way: low temperatures encourage the vegetable cells to store up a higher concentration of sugars and starches rather than water which would be more readily lost, leading to undesirable shrinkage.

If possible, it's better to harvest vegetables in dry weather than during a rainy spell. While the ground may be more difficult to dig if it is hard, especially if the humus content is low, vegetables dug during dry weather may keep better because (1) they are not plumped up with recently absorbed water, and (2) they have less soil clinging to them and consequently need less handling to prepare them for packing away. Whether the soil is dry or wet, we usually use a fork rather than a shovel to unearth our vegetables, because the fork is less likely to slice through the roots.

One of our country lore consultants tells about stopping to watch a farmer and his family pulling cabbage from their field and loading it in a cart in a steady rain. "Why," he asked, "don't you wait until tomorrow when the sun will shine?"

The farmer seemed to know what he was doing. Continuing down the row, he said, over his shoulder, "The moon sign's right today. The cabbage'll keep better." (We might also add that in another day's time the large heads might have split after absorbing the rainwater.)

Just as many gardeners try to plant seed in a favorable phase of the moon, many moon sign devotees are convinced that produce picked in the right sign of the moon will keep better.

*The Foxfire Book, The Moon Sign Book,* our current *Farmer's Almanac,* and local folklore all seem to agree: if you pick apples and pull root crops during the decrease of the moon, in the third and fourth quarters, bruised spots will dry instead of rotting and the food will keep better.

Grain to be stored, they say, should be reaped just after a full moon.

We leave it to you to decide whether or how to use this information. We have not yet experimented with harvesting by moon sign ourselves, but we are less skeptical about it than we once were.

*Cleaning* should not be necessary in most cases. It's perfectly all right to leave a light coating of dusty soil on the surface of your root vegetables. Gently brush off excess dirt, but avoid scrubbing or washing the roots. They'll keep better if you don't clean them up too vigorously. You can scrub them well just before you eat them. When it is necessary to dig the vegetables from wet ground, let them dry off just a bit (but keep them cold) so that you can carefully knock the clods of heavy soil from their surfaces.

*Handling* should be gentle. Never toss the vegetables into a cart. Always put them down with deliberate care. They'll last much longer if you do. Bruising at harvesttime can cause a lot of unnecessary spoilage. The bruised flesh will be the first to rot in storage. There are bacteria and fungi everywhere looking for just that sort of terrain on which to live.

*Selection* will prolong the life of your stored hoard, too. Poorly developed, nicked, or bruised fruits and vegetables aren't likely to make it through the winter. In spoiling, they may encourage decay in other sound foods touching them. Store only your best produce — sound, mature, well developed. Set aside the immature and cut-open or insect-damaged specimens to make a big pot of stew, soup, or mixed pickles, or use them to fatten a pig.

*Curing* isn't necessary for most root vegetables. They should, in fact, be hustled quickly from their cold moist underground situation to cold, moist storage. Most root vegetables should not be exposed to the drying sun after digging. (Exceptions are noted in chapter 8.) Bulbs like onions and garlic need a week in the sun to dry for storage, and squash and pumpkins should be exposed to sun for about two weeks to develop a hard rind. Sweet potatoes should be cured before storage too. Those vegetables that do need curing will not keep as well if you skip this process.

*Clipping* tops off should be done right after digging, before packing the vegetables away. Why not leave the leafy tops on? For one thing, they'll draw off a good bit of moisture from the root as they wilt. And they'll turn slimy sooner or later when the vegetables are packed away, if indeed you are able to get them decently packed up with all those extra leaves in the way. Replanting root vegetables indoors in soil or sand especially for production of tender new leaves is another subject, covered in chapter 10.

Leafy tops of parsnips and beets are good to eat. When time permits, I like to have a big pot of vegetable soup simmering on the wood stove during harvest days, all ready to absorb the extra rib of celery, leaf of broccoli, runty head of cabbage, and sliced-open potato that are too good to waste. Don't get carried away with trimming off extraneous parts of your vegetables, though. Root tips and small feeder roots should not be broken or pulled off to neaten things up. Even small skin breaks invite spoilage. Wait until serving time to trim those odd knobby parts, forked roots, and excess rootlets.

# Susceptibility of Growing Vegetable Plants in Garden to Frost Damage

**Very Susceptible**

| | |
|---|---|
| Cucumbers | Eggplant |
| Lettuce | Squash |
| Sweet peppers | Sweet potatoes |
| Tomatoes | Pumpkins |

**Moderately Susceptible**

| | |
|---|---|
| Beets | Onions |
| Broccoli | Celery |
| Young cabbage | Spinach |
| Carrots | Parsley |
| Cauliflower | Peas |
| Escarole | Radishes |
| Garlic | |

**Least Susceptible**

| | |
|---|---|
| Kale | Parsnips |
| Collards | Brussels sprouts |
| Kohlrabi | Mature cabbage |
| Salsify | Turnips |

# 5

# Life After Picking

Very often I stop and think (one must) of the hidden life in the cellar storeroom during the winter. I think of all those beings taken from the garden, whom the plucking has not quite killed, and who, without light, barely breathing, live like fakirs, or in limbo.

Fernand Lequenne
*My Friend the Garden*

If it seems odd to you to talk about how vegetables breathe, you probably have plenty of company. It's commonly assumed, after all, that fruits and vegetables are no longer alive when they're removed from the plant that produced them. Vegetables and fruits continue to live, however, even after they're picked. The foods you bring in from the garden are still breathing while you hold them in your hands. If you're counting on those foods to nourish you over the winter, then, it might be a good idea to consider how their continued respiration affects their keeping and eating qualities, and what you can do to control their life processes.

We've said that harvested vegetables and fruits breathe. Their respiration, like ours, involves taking in oxygen from the air, combining that oxygen with substances in their flesh, and then giving off carbon dioxide, water vapor, and heat. Perhaps at this point you find yourself picturing those apples and onions in your cellar huffing away, breathing and puffing and even conspiring to get out of the cellar. No one has ever *seen* an onion or a cabbage breathe, of course, because they take in oxygen through their pores, but the results of respiration — the intake of oxygen and the release of carbon dioxide, water vapor, and heat — have been *measured* in laboratories.

All the while your produce is breathing — and it will continue to breathe until you cook it or eat it, or until it rots — it is changing. Becoming mature. Ripening. The faster the rate of respiration, the more quickly the piece of food will age and deteriorate. Respiration uses up natural sugars in the foods, which is why vegetables sometimes change in flavor after long storage. Overripe, over-mature vegetables are less appealing to eat and they spoil more quickly.

The aging process is a natural one, of course, and it can't be stopped. Each vegetable has its own internal time clock. Melons, tomatoes, and berries need a ripening period to attain their full flavor, sweetness, and best texture. These foods may be ripened in storage if picked immature, but must be used fairly soon after harvest. Carrots, beets, potatoes, squash, and most other vegetables don't ripen, or improve, after picking, but they may be kept on "hold" for a much longer time.

You can see that it is to your advantage, if you'd rather not have to run to the store for carrots in January, to pack away vegetables under conditions that will retard the normal aging process and slow respiration. Lowering the temperature is the single most important thing you can do to promote vegetable longevity. Cold not only slows respiration, it also counteracts the emission of heat by the vegetable, which is a side effect of respiration. Cold is so important that increasing storage temperature by ten degrees will drastically reduce the usable life of your vegetables (see chapter 8 for exact temperature ranges preferred by different vegetables). At higher temperatures, fruits and vegetables have a fast rate of respiration, ripening, aging, and deterioration.

Fruits and vegetables contain a lot of water. Most are 80 to 90 percent water by weight, but they lose water from their tissues during storage. When water loss is marked, the foods shrivel and lose their appealing texture and some of their vitamin C. A small amount of water is given off in respiration in a complex chemical interchange between the food and the air around it. Much more tissue water is removed, though, by evaporation and transpiration. Evaporated moisture is lost from the surface of the food. Water lost by transpiration passes from the internal tissue of the vegetable through the pores and leaf surfaces into the drier surrounding air.

Water loss is highest for most fruits and vegetables during the period immediately after the harvest. Raising the humidity helps to reduce water loss by transpiration, but if the vegetable retains heat from the earth or from postharvest exposure to the sun, it will continue to lose tissue moisture even in 100 percent humidity. You can see, then, that rapid cooling of storage vegetables helps to maintain their high quality, and that *keeping* the storage temperature cold also helps. Because warm air can hold more water vapor than cool air, it will absorb more moisture from stored food than cold air, and water has a greater tendency to evaporate as the temperature rises.

Some cold-sensitive vegetables like sweet potatoes and peppers may suffer chilling injury from low but not freezing temperatures in storage. Chilling injury is not always immediately apparent, but it usually shows up when the vegetable is brought into a warmer place, in the form of pitting — black or corky spots in the flesh. The majority of root vegetables like to be kept as close as possible to 32 degrees F.

Some fruits and vegetables release the plant hormone ethylene gas — among them apples, tomatoes, muskmelons, pears, plums, and peaches. Ethylene gas can affect the quality of other stored vegetables. It tends to hasten the aging process, resulting in sprouting in potatoes, pale color in peppers, cucumbers, and leafy vegetables that should be green, and bitter flavor in carrots. Keeping the temperature below 45 degrees F helps to minimize ethylene's aging effects. If you have many bushels of apples and pears, you may need to keep them separate from your vegetables, but usually a few baskets of each can coexist without serious problems.

The great majority of the vegetables we store for the winter are biennials — plants that form seed during their second growing year. Onions, leeks, the cabbage family, and root crops like carrots, beets, parsnips, and salsify are all biennials. Nature *intends* biennials to keep well so they can go on to bloom and bear seed after laying low all winter. It's not just our idea, then, to keep these vegetables in good shape until spring. Winter survival is in their genes. We're not outwitting natural laws, but cooperating with them, when we build root cellars and dig storage pits. Now that you know a bit more about how these vegetables behave after picking, you're in a better position to take good care of them, so they can take good care of you.

# 6

# Spoilage

Grandpa Snedd . . . was the saving kind. When he sent a youngster downstairs for a bowl of apples for the evening snack, he'd order, "Look through for any that have spots, and bring those. One bad apple can spoil the whole crate."

Aunt Carrie was his daughter, but she never agreed. "Bring the best apples you can find," she'd say. "That way, even when they're all spotted, we'll still be getting the best of the batch."

Eupha Shanly
"The Farm Cellar," *Poor Joe's Pennsylvania Almanack*

We've become so accustomed to controlled standardization in our purchased foods, to at least surface perfection, that there is no room in our scheme of things for the idea of decay. When you buy two cello-packed green peppers in the supermarket, you expect them to be sound, and that's reasonable. (What most of us fail to realize is that fully 25 percent of the produce harvested commercially in the United States is never eaten because it spoils after harvest but before it can be purchased. Mechanical handling and long-distance shipping of produce that was picked green no doubt contribute to this unfortunate waste. Surely you and I can do better than that.) If you've frozen or canned foods carefully, you'll get little if any spoilage, and even many home gardeners consider a rotten vegetable to be a sign of failure.

Not so, though. The bad apple and the squishy squash happen to all of us who store foods unprocessed, even under the best of conditions. It's natural for a few vegetables to deteriorate before winter is out. The idea in root cellaring is to grow enough, and store enough, so that a few small losses won't matter. Don't be surprised or threatened when one of your pumpkins collapses from within or an apple goes leathery with brown rot. Expect some spoilage and work around it.

If one-third or more of your stored beets or apples or whatever fails to make it into serious winter storage, though, then it's time to look for reasons. Were the vegetables immature? Bruised? Washed, or too closely trimmed before storage? Were storage temperature and humidity right for the food? Were they stacked in large poorly ventilated piles?

When a small percentage of a stored crop decays in storage, you'll often find that it is the bruised sweet potato, the immature squash, the stemless pumpkin that goes first. Bacteria and fungi take advantage of small skin breaks and poorly cured innards to set up thriving colonies that continue to break down the unsound tissue. While this is annoying when it happens to food you intended to eat, consider that we'd be stuck with mountains of useless garbage if it were not for this action of micro-organisms, the ultimate reducers.

Most serious gardeners have great respect for the process of decay, which turns a heap of leaves and plant trimmings into rich, crumbly compost. So if you must toss a few vegetables on the compost pile, just figure that they started their journey in that direction a bit early. You'll meet them again in next spring's compost and next summer's garden produce.

Deterioration of stored vegetables is not always absolute, though. Often you can catch signs of decay — small black or soft spots, weeping areas, and so on before they spread. If you cut out all the bad parts, you can still use the food. But be sure to avoid including any moldy tissue with the parts of the vegetable you intend to use. Many fungi and bacteria produce toxins as by-products of metabolism, and you don't want to eat these toxins with your supper.

A half-gone squash you may not consider sound enough to rescue for the table may still serve an intermediate purpose before sinking back into the earth. Give it to the chickens or cut it into pieces to recover the seeds, which usually remain in good condition for a while after the flesh decays.

Some people use up unsound foods first, in order to catch them before they spoil. This frugal practice makes a good deal of sense at the beginning of the storage season, but if you find as winter progresses that you are always eating the second-rate vegetables so they won't go bad on you, and leaving the best ones in the bin, then perhaps you should reconsider priorities. As long as food is not actually scarce, enjoy the

Rotten squash provide an afternoon snack for Petunia the pig.

good stuff while it's really good. If a handful of questionable vegetables is nagging at your conscience, pickle them or make apple butter, squash butter, sauerkraut, or catsup if you have the time. If not, discard them unmourned and concentrate on the good apples in the barrel.

Occasionally some produce will freeze, because it was either left outside on a very cold night or insufficiently insulated in storage. All is not lost. Carrots, beets, and other sweet root vegetables may be used in soups and stews even if they've been frozen. Frozen onions and cabbage are perfectly all right to use too. Once frozen and then thawed, the softened tissues of the vegetables spoil much more readily, though, so frozen vegetables that have thawed should be used promptly.

The only way to prevent the spread of decay from unsound vegetables to good ones is to inspect your stored hoard weekly and cull the early quitters from your good keepers. There are 50 kinds of rot that can affect a stored apple and all of them are ready and eager to go to work on all the good apples that are in contact with a single bad one.

These weekly checking sessions can be just a chore, or they can give you a chance to gloat (while shivering slightly) over all the good food you've managed to grow and put by. You choose.

# 7

# Food Value in Winter Keepers

Each season offers a chance to improve garden tilth and plant food content.

R. Milton Carleton
*Vegetables for Today's Gardens*

Vegetables are valuable for their content of vitamins, minerals, and fiber. How well do these important components hold up in storage?

The vitamin A content of squash (and presumably pumpkin) actually increases during storage. The variety grown and the conditions during storage determine the amount of the increase, so no exact figures can be given.

B vitamins are apparently quite stable in storage. Losses are usually small or negligible. Growing conditions can affect the content of certain B vitamins, though. For example, some studies have shown that application of trace minerals to the growing plant increases both thiamine and riboflavin content of the food. And in another study, turnips harvested in the morning contained more riboflavin than those harvested later.

Fiber and mineral content remain essentially unchanged in sound vegetables. Some vegetables become more fibrous as they age.

When green vegetables wilt, they lose vitamin C content. Well-fertilized kale, cabbage, and collards contain more vitamin C than those grown in low-nitrogen soil.

# SECTION THREE

## All the Winter Keepers and How to Treat Them

# Storage Requirements of Vegetables and Fruits

**Cold and Very Moist** (32–40 degrees F and 90–95 percent relative humidity)

Carrots
Beets
Parsnips
Rutabagas
Turnips
Celery
Chinese cabbage
Celeriac
Salsify
Scorzonera
Winter radishes
Kohlrabi
Leeks
Collards
Broccoli (short-term)
Brussels sprouts (short-term)
Horseradish
Jerusalem artichokes
Hamburg-rooted parsley

**Cold and Moist** (32–40 degrees F and 80–90 percent relative humidity)

Potatoes
Cabbage
Cauliflower (short-term)
Apples
Grapes (40 degrees F)
Oranges
Pears
Quince
Endive, escarole
Grapefruit

**Cool and Moist** (40–50 degrees F and 85–90 percent relative humidity)

Cucumbers
Sweet peppers (45–55 degrees F)
Cantalope
Watermelon
Eggplant (50–60 degrees F)
Ripe tomatoes

**Cool and Dry** (35–40 degrees F and 60–70 percent relative humidity)

Garlic
Onions
Green soybeans in the pod (short-term)

**Moderately Warm and Dry** (50–60 degrees F and 60–70 percent relative humidity)

Dry hot peppers
Pumpkins
Winter squash
Sweet potatoes
Green tomatoes (up to 70 degrees F is OK)

# 8

# Vegetables

I know of nothing that makes one feel more complacent . . . than to have vegetables from his own garden. . . . It's a kind of declaration of independence.

Charles Dudley Warner
*My Summer in a Garden*

Good winter keepers belong to one of several categories, depending on how much moisture and cold they need to stay in good condition. (See chart on page 51.) But they have their individual quirks, too, and their special qualities, which should be noted. Here, then, is a rundown on all the vegetables you're likely to keep, with information on when to harvest them, how to prepare them for the root cellar, what storage conditions they need, and approximately how long they'll last.

You'll want to notice which vegetables are most likely to be damaged by frost, either outside in the row or in storage. A low temperature of 28 to 31 degrees F is considered to be a light frost; moderate frost is 24 to 28 degrees; and severe frost is anything below 24 degrees. You also will note that, although a certain ideal minimum temperature is recommended, in practice you can often keep a generous supply of vegetables in good condition for a while at somewhat higher temperatures, although they won't last as long as they would in a colder place. It's good to know the ideal and to aim for it, but don't give up if your root cellar won't go below 40 degrees. You can still store lots of good food there.

**Beets** Beets are good keepers. While they don't always last quite as long in storage as carrots and potatoes, they've kept for us well into March, providing color and nutrition on our winter dinner plates. Beets should be harvested before carrots and potatoes, both because they're a bit

These Long Season storage beets look big and tough, but actually are tender and delicious.

more susceptible to frost damage and because of their growing habit; often the roots are at least half out of the ground by midfall, openly exposed to frost unless well mulched. Pick mature beets well before killing frost, then. Whack off the tops, leaving a one-inch stub of stems. If you trim the beet leaves too close to the root top they'll bleed, and then they'll deteriorate faster. Don't cut off the pointed tip of the root. Make good use of those vitamin-rich leaves by serving some fresh-cooked and chopping and freezing the rest, or dry them and crumble them into your winter soups.

Pack the beets in layers of damp sawdust, moss, or sand and keep them cold and very moist — as close as possible to 32 degrees F and 90 to 95 percent humidity. Leaves or plastic bags will do for storage if you can't obtain sawdust or sand. (We prefer sawdust.) Count on a life of at least two months for stored beets. When well kept, they'll often last a good four to five months.

**Broccoli** Broccoli is at its best when the flower buds are still tightly closed, firm, and a definite green. Yellowed or open buds, or those actually in bloom, will be several notches lower in flavor and will not last as long in storage. (Purple varieties of broccoli like Royal Purple and others should, of course, be their usual dusky violet color, not green or yellow.)

Late broccoli crops are often less affected by insects when mature than early plants. Broccoli is moderately susceptible to frost. It will survive the first few mild fall frosts but should be harvested before severe frost. Cut the central head and side florets in the cool of the evening, if possible. Trim off a panful of the tender side leaves for a dish of greens for dinner, then bag each head in plastic and keep it cold and very moist — as close as possible to 32 degrees F and 95 percent humidity. The heads will remain in good shape for about a week, sometimes as long as two weeks if conditions are just right. (If you freeze broccoli for winter meals, freeze your fall crop, not your summer broccoli. Broccoli has better quality in cold weather and it's sensible not to tie up freezer space unnecessarily when a crop can be grown later.)

**Brussels Sprouts** A flourishing Brussels sprout plant, its stalk studded with close-packed firm leafy knobs, is a heartening sight in the November garden when most other fresh produce has come indoors or gone underground. Brussels sprouts are unusually frost-resistant, and though they still can't match kale for hardiness, we count on them for fresh table fare at least until Thanksgiving, and often beyond. (Unless the sheep get out, as they did this year. When they do, they head straight for the Brussels sprouts row. Sheep aren't so dumb. After Brussels sprouts, they polish off the cauliflower leaves.)

If you're growing sprouts for the first time, wait to harvest them until they've endured several frosty nights. Frost improves their flavor dramatically. Unfrosted sprouts taste a bit coarse, but when frosted their flavor is delicious. Firm, compact sprouts about one inch in diameter

Brussels spouts thrive in the garden through November, and several frosts will improve their flavor.

with good green outer leaves are tops in quality and your best bet for storage. Sprouts you've picked off the plant will keep in perforated plastic bags for three to five weeks, depending on how closely you can approximate their ideal: cold (near 32 degrees F) and damp (90 to 95 percent humidity).

If you're very ambitious and have plenty of room, you could even dig up a plant or two before the ground has frozen hard and replant it in your root cellar. The root system is fairly shallow and compact and should fit in an ordinary household bucket or tub. Such a ploy would keep two people supplied with fresh sprouts for perhaps an extra month. Or, you could simply hang the bare-root stalk full of sprouts in your root cellar.

**Cabbage** For your winter supply, pick crisp solid cabbage heads that seem heavy for their size. Studies reported by Pennsylvania State University's Cooperative Extension Service reveal that the perfect storage cabbage is one that is picked just before the top wrapping leaves lose their bright green color. Technically, such a head is just the slightest bit short of full maturity. Heads that are almost but not quite fully mature, they've found, will retain their green color for a longer time.

Heads should be firm and solid. Leafy, loose-wrapped cabbage heads don't keep well. These should be used promptly as fresh vegetables or processed into sauerkraut or pickled relish. In practice, because it's not always possible to time the harvest so precisely, I'd choose solid if fully mature heads over soft, leafy ones for storage. Pull the heads from the ground, root and all. Trim off all the loose outer leaves so that air can circulate freely around the heads when they're put away. When air circulation is poor, fruits and vegetables stored close to cabbage may absorb an undesirable flavor. If your root cellar is not as well ventilated as you might wish, you may want to store your cabbage (and its relatives, broccoli, collards, Brussels sprouts, and so on) separately.

Cabbages showing signs of disease should not be put into storage because such diseases spread quickly in cold damp places. At temperatures above 35 degrees F, two common storage diseases of cabbage — Botrytis leaf mold (a gray mold) and Alternaria leaf spot — may affect bruised or nicked cabbages. Moral: store sound heads and keep them cool: 32 to 40 degrees F, and about 90 percent humidity.

Cabbage releases a lot of moisture in storage, especially in fall and spring. It also breathes out cabbagey fumes, which can make your house smell like a ship's hold. You might find that you need to plan for long-term cabbage storage somewhere other than your basement — in a shed, outdoor underground root cellar, barrel, trench, or clamp. (See chapter 12.) The amount of ventilation in your basement will largely determine whether you can keep cabbage in your house.

In the root cellar, you can simply place the heads on shelves, spacing them several inches apart; you can tie a string around the root and suspend the cabbage head down from a rafter; or you can layer heads in hay or wrap them individually in newspaper. No single method will be ideal for everyone. Try several methods to see which seems best for you. Should some cabbage freeze, it may be cooked when thawed but it will

Chinese cabbage is one of the few leafy vegetables to survive low temperatures and store well for extended periods of time.

This Chinese cabbage has been dug up and replanted in a bucket for storage.

not keep well raw if it thaws after freezing and should not be allowed to refreeze.

You might very well get away with keeping cabbage in your basement if you wrap the heads in several thicknesses of newspaper or spread the heads, root side up, on a two-inch bed of dirt near the cellar wall, and cover them with more dirt.

**Cabbage, Chinese** Choose mature, solid heads for storage. Be sure to pick them before severe frost, but they'll endure many mild freezes. In fact, frost seems to improve their flavor. Chinese cabbage keeps quite well for such a leafy vegetable. We used the last head in late February last year, from a November harvest, and it was still in good shape inside. The outer layer of leaves wilts, but the interior remains crisp and leafy. The old way with Chinese cabbage was to arrange the heads in shallow piles, heap hay over the piles, and toss on soil or old boards to hold the hay down. Such piles will keep up to four weeks in the garden, and at least a month or two under cover.

We prefer to lift the plants root and all, pull off the flouncy outer leaves and replant the heads close together in a box of soil in the root cellar. If the plants aren't stored upright, they become lopsided in appearance, but they taste the same. Water the roots as you add them to the box, and water again once or twice during the winter as necessary. Try to keep water out of the crown of the plant, though, or the leaves may rot. Thirty-two to forty degrees F and 90 to 95 percent humidity provide ideal conditions for Chinese cabbage.

We store Chinese cabbage in the cellar under our living room, by the way, and it never smells "cabbagey" — not even up close. When a cabbage rots in storage, though, that *does* smell. Such specimens should be shovelled out promptly.

**Cardoon** This domesticated member of the thistle family is an herbaceous perennial. In mild climates it sets seed, dies back to the ground, and comes up from the root the following year. It also self-sows freely. Some horticulturists theorize that this plant could have been the first intentionally cultivated crop. Artifacts found with buried cardoon seeds indicate that the seeds had been gathered more than 30,000 years ago. If that is true, then present-day devotees of cardoon are heirs to an ancient practice. Let's hope that gardeners will continue to grow unusual vegetables like this one. We are all poorer when good old varieties, representing the gene pool of thousands of years of seed-saving, are allowed to die out.

A cardoon plant looks like an enormous bunch of celery. The leaves are bitter and the outer stalk tough, but the inner stalks and the core, which is crisp and meaty like that of celery, are tasty and pleasantly bittersweet. Although cardoon seeds need warmth to germinate, the mature plants can withstand temperatures in the low 30s or even high 20s. The cardoon is customarily wrapped near the top and banked with earth on both sides to blanch it in early fall, a procedure that helps to give it a milder flavor and protect it from frost. Real cardoon enthusiasts have learned to go one step further to prolong the season: they dig foot-deep trenches alongside the plant, gently bend the stalks over into the trench, taking care not to snap off the taproot, and cover them with a foot of earth, leaving a tuft of leaves exposed. In all but very wet weather this will ensure another two or three weeks of cardoon harvest. The cut stalks and heart, trimmed of their fuzzy, prickly leaves, will last for a week or two in a cold place if kept in plastic bags. Our 1895 gardening book advises, "Take up before frost and store like celery." Sounds like an heroic task, unless you slice off a lot of leaves. Those plants are big.

**Carrots** The reliable carrot is the backbone of any food storage plan. These mild, tasty, vitamin A-rich roots ae easy to grow and they'll last until May or beyond when kept cold and moist — 32 to 40 degrees F and 90 to 95 percent humidity. Carrots with a high fiber content keep best. Those grown in heavy soils sometimes become quite fibrous in

Packing carrots in sawdust in a carton is our favorite storage method.

struggling to extend themselves into the close-packed ground. Such carrots won't win any flavor awards but they may still be solid next June. We've stored carrots from our early spring planting, but find that the later planting in June retains better flavor in storage.

Dig carrots before the ground freezes hard. Brush off any dry dirt that clings to them and break off the tops. Some gardeners leave a half-inch stub of leaf stem, but we usually break the stems off right where they join the root top.

Our favorite carrot storage method is simple: Spread an inch-thick layer of damp sawdust in a carton, box, or can. Arrange the carrots side by side on this bedding. They can touch, but should be kept in a single layer, not piled on each other. Put a one-inch layer of damp sawdust over the first course of carrots. Press another tier of lined-up carrots into that and continue to alternate carrots and sawdust until the container is full, ending with a covering of damp sawdust. If your storage area is drier than you would like it to be, cover the carton with damp newspaper. If you have no sawdust, you can use damp sand, bagged peat moss, wild moss, or leaves.

Carrots also keep well right in the garden row. Spread a one-foot to two-foot-thick layer of mulch over the row before the ground freezes hard and you'll be able to dig up fresh carrots for another three weeks or so, and in January thaws if snow doesn't cover the ground. Mice sometimes tunnel under mulch and eat the carrots. When this has happened in our garden, we've found a row of neat carrot-shaped holes at spring digging time. The solution is to put down a layer of hardware cloth or screening over the row before mulching.

**Cauliflower** This cabbage relative thrives in cool weather but severe frost will spoil its keeping quality. Since it keeps only a short time at best, you must weigh the potential loss of the crop to frost against the hazards of picking too early before the root cellar is cool enough to hold the heads in good condition. Some years we hit it right, other times we don't, but even if the heads freeze, they may still be cooked (very briefly) and served fresh. Last year when some of our cauliflower froze in the fall we were able to keep half a bushel of the frozen heads on the shady porch for a week before they thawed. If warm weather had followed the freeze, this would not have been possible.

Cauliflower heads are ready to pick two to three weeks after you've blanched them (if blanching was needed) at the three-inch stage. Once they've formed a solid head, they won't grow any larger. Instead, the florets that make up the "curds" on the head will begin to bloom, resulting in a ricey texture.

Store only good solid cauliflower. For root cellar storage, pull cauliflower up by the roots and either replant in sand, hang by the root, or spread on shelves. The heads will stay crisper if you enclose them in plastic bags. Cauliflower will last from two to four weeks when kept cold (32 to 40 degrees F) and moist (90 to 95 percent humidity). Use it soon, because its storage life is short.

**Celeriac** These knobby celery-flavored roots are turnip-size, but not turnip-smooth. Masses of small tangled surface roots give them a gnarled appearance. You can snip off the longer, finer roots but don't cut too close to the main body of the vegetable. Shake off loose dirt. Trim the tops to a one-inch stub, then spread the roots out on slatted shelves. Or, if you have a large crop and expect to store them throughout the winter, cover the roots with damp sawdust, moss, or sand. Like most root crops, celeriac keeps best when cold and moist: 32 to 40 degrees F and 90 to 95 percent humidity. Celeriac planted in early spring may turn woody by fall. Put in a special mid-or late-spring planting for winter storage. Celeriac keeps well — it's a vegetable you can count on.

**Celery** Celery that matures in the fall may often be stored right in the garden for a month or two, then brought into the cold cellar for another month or two of storage during the coldest winter months. If you have a good compact row of celery, you can gradually bank up the plants on both sides with loose soil. The plants should be covered up to their leafy tops by the time hard frost hits. Then, for further protection, pile ten inches of straw or hay over the row, topped with boards to keep it in place.

Or, you can dig a trench two feet deep and about a foot wide, dig up the celery plants, and replant them in the trench. Try to keep a good ball of soil around the roots when you move the plants. Water the roots when you set them in place, but avoid spraying the leaves. To improvise a roof over the trench, the United States Department of Agriculture suggests propping an upright board at the edge of the trench and banking soil on a slant up to the board, then covering the sloping soil bank with a thatching of cornstalks, scrap boards, and such. Good drainage is essential for trench-stored celery. It will rot if waterlogged. Some gardeners arrange bales of hay around and over their celery rows. This method can work quite well if you get the hay on before the ground freezes hard.

In the root cellar, celery keeps well at 32 to 40 degrees F and 90 to 95 percent humidity. Dig the roots and replant them in the cellar in boxes of sand or soil, packing the plants close together in the box. Use a spouted watering can to keep the roots moist — if the leaves and stalks get wet, they may rot. Handle celery carefully when harvesting, as bruised stalks are quite prone to storage diseases. Cabbage and turnips should be kept separately or the celery may absorb some of their flavor. Well-kept, home-grown celery is worth the effort it takes to keep it crisp and fresh for winter.

**Chives** Chives are not really a root cellar crop, because they need light to keep growing, but they are one of the easiest of all green growing things to bring indoors to tide you over the winter. While you're digging up roots, spade up a healthy clump of chives, too. Perhaps they need thinning anyway. Plant the clump in your best potting soil with an inch of drainage pebbles lining the bottom of the pot and keep the pot on a sunny windowsill. Replant the chives in early spring if you wish, after snipping its mild onion-flavored quills all winter to garnish your hearty root dishes.

**Collards** Not for the long run, but harvested collards will fill in with a week, or two or three, of good fresh produce after the rest of the garden has gone to bed for the winter. The plants are very cold-hardy and even in the north they'll often stand well right in the garden until well after Thanksgiving. In the South, collards often stay in the garden all winter. What could be easier! Frost improves their flavor, but when black frost or severe snow threatens your northern-grown collards, pick the leaves or pull the whole plants. Leaves in perforated plastic bags will keep well in a cold, damp root cellar (32 to 40 degrees F, 90 to 95 percent humidity) for a week or two. Stack whole plants and cover them with a damp feed bag or old towels.

**Cucumbers** What's this, cukes in the root cellar? Well, not for long, but have you ever noticed how many gardeners keep a basket of cucumbers waiting on the porch until they have a chance to use them? The cukes will keep longer in a cool, rather damp place. In early fall, when the vines are yielding their last fruit, the root cellar may not be cold enough for root crop storage, but chances are it's just right for keeping cucumbers, which do best at 45 to 55 degrees F and 80 to 90 percent humidity. Remember that your winter fare will be more varied for every week that you can extend the season, so it's worth keeping cukes as long as you can. If not overmature (yellow), they'll last up to two to three weeks under good conditions. You might try cooling the cucumbers in cool well water before putting them in the root cellar. After they're dry, they can be put in perforated plastic bags if you wish. Note, though, that they need less moisture than root crops.

**Eggplant** Eggplant isn't a long-term storage vegetable either, but it will *keep* for about a week or two at around 50 to 60 degrees F and 80 to 90 percent humidity. Temperatures lower than 50 degrees will overchill the eggplant and cause spoilage. Eggplant will stay in good condition as long in the root cellar as in the refrigerator, perhaps longer, so save your frig space for milk and meat. The day you stash it away, delve into your recipe collection for ten ways to serve eggplant, and use it promptly while it still has that garden glow.

**Endive and Escarole** These tender, full-flavored leafy greens are an excellent crop for continuing fresh salads into winter. For the most tender, mild-flavored leaves, blanch the plants by tying the leaves together in late September. (This taste preference is a matter of habit, though. Many people have learned to prefer the more nutritious unblanched green leaves.) Endive and escarole stand well in the garden through mild and moderate fall frosts, with some light covering. We dig the remaining plants in November and replant them close together in boxes of soil or sand in the cold cellar. The first year we stored escarole, we left the leaves tied up, but found that a lot of the plants turned mushy. They seem to do better for us if the heads are untied when they're brought in, to allow for more ventilation. Then the leaves don't rot, although they do take up more room in the box.

**Garlic** Harvest garlic when the tops have died back to the ground. Don't wait too long, though. Garlic tops rot more quickly than onion tops and the row can be hard to find if the tops disintegrate and weeds grow tall. You'll need to dig the bulbs; the tops aren't usually strong enough to pull. The bulbs will keep all right without tops, but if all the tops are gone you won't have any to braid. Shake off the loose earth and cure the garlic bulbs in the sun for several days to harden their skins. Clip off the bunch of roots right up close to the bulb, leaving as short a stub as possible. Also snip off the tops unless you intend to make garlic braids. We usually set aside a bunch with nice tops and large bulbs to make several braids.

To make a garlic braid, tie together a length of strong white string and the dried ends of three garlic tops of different lengths. Then braid the tops together, carrying the string along with one strand to strengthen the braid. When one dried garlic top has been braided down to the bulb, start to braid in another one to take its place. When the braid is as long as you want it to be, tie a string around it at the bottom, attach a hanging loop to the top, and clip off the tip ends of any dried tops that stick out.

Garlic does not keep well in the refrigerator, as you may already have learned to your dismay. It's too damp in there, and the bulbs rot. In a warm, dry spot the bulbs will shrivel, but cheer up! There *is* a good way to keep garlic: 35 to 40 degrees F at 60 to 70 percent humidity is perfect. We keep ours over winter in paper bags in a cold back room and it lasts until April and beyond, from the previous August when we harvest it. We've also successfully kept garlic in net bags hung over the basement stairway. It doesn't take much room and doesn't smell until cut, so you can even keep it in the guest room or attic if temperature and humidity are right. When we first bring the garlic in, we often put it unwrapped in the cool basement until fall when our unheated rooms start to cool off.

**Horseradish** Horseradish is very hardy and keeps quite well. It's a good thing it's at its best just when we need it most: in fall to go with the newly butchered pork and in spring to season the hard-boiled eggs when the hens resume laying. Horseradish roots seem to need a winter freeze and don't do as well in the warmer parts of the South.

The horseradish root makes its prime growth in the fall when weather turns cool. Roots may be dug from unfrozen ground any time after September. To prolong the harvest, you can cover the patch with bales of hay so the ground will stay soft enough to dig for another few weeks. If you want to be sure of having fresh horseradish over winter even if the garden is drifted deep in snow, dig up large roots in late fall, just before the ground freezes hard. Store them in damp sand or sawdust in a cold, damp place — 32 to 40 degrees F and 90 to 95 percent humidity. Or you could keep them in the cold frame, layered in leaves, until you need them.

When the ground thaws in early spring, you'll find the roots you left in the patch still in good condition, and they'll remain good for grating until they start to sprout new top growth. (Hint: If you replant the top inch of the horseradish root after cutting off the rest of the root to grate, you'll often get a new plant.)

**Jerusalem Artichoke** The sunchoke, as the Indians called it, is another no-processing vegetable that keeps very well indeed right in the ground. The tubers live over the winter and sprout a new crop the next year. You never need to worry about harvesting all the Jerusalem artichokes and leaving none for next year; there are *always* nubbins with eyes and missed tubers left in the ground after even the most thorough digging. We enjoy sunchokes as a late-fall vegetable, starting to dig them in October and continuing until the ground freezes over. We toss hay over the bed to keep the soil from freezing hard so we can sneak in an extra month or so of fresh artichokes. Sometimes mice will tunnel under the hay and nibble on the tubers if they're close to the ground surface. To

Digging Jerusalem artichokes in early spring.

discourage mice, cover protruding tubers with another inch or two of soil.

Once dug, artichokes gradually shrivel on exposure to air because their skins are so thin. They'll keep in a cold place for several weeks in plastic bags and for a month or two in damp sand in a cold root cellar or garage.

**Kale** This sturdy, leafy green cabbage cousin is a super vegetable — high in vitamin content, extra hardy, and easy to grow. You can depend on kale for good eating in the fall when most other green things have faded away. The flavor is, in fact, at its best after frost. When the ground is ringing hard, you can still harvest kale, if you've planted enough of it. Even in the dead of winter, kale will remain in usable condition in the garden, if protected by a blanket of mulch. Don't smother the crowns of the plant, just tuck mulch up closely around the stem and toss some hay lightly over the crown in severe climates. Then you can just brush the snow away and snap off a few leaves for dinner. *That's* vegetable independence!

**Kohlrabi** Store only the bulbs from summer planting, as spring-planted kohlrabi bulbs will be tough and woody by fall. Bulbs larger than three inches also are likely to be tough. Small, quickly grown bulbs should be crisp and tender. Trim off the leaves (they're good to eat) and cut off the root. When kept cold and moist (32 to 40 degrees F, 90 to 95 percent humidity), kohlrabi will remain in good condition well into the winter if packed in damp sand or sawdust. It doesn't last as long as carrots and potatoes, though, so use it and enjoy it while the year is still young.

**Leeks** Dashing out in the December dusk to snatch a thick green leek from our windblown garden, I reflect on the convenience of having a kitchen garden as I bear my prize inside where it's warm and light. Leeks will wait for you all fall, and they don't grow tough and stringy in the process. You can even leave them in the garden row all winter. When well mulched or banked with soil, at least half the crop will come through the winter and we've even had several leeks in a bare-ground planting survive a snowy winter. To be sure of having leeks when you want them all winter, dig some up root and all and replant them in your root storage area in a box or tub of sand or soil. They do very well for us at around 42 degrees F. A lower temperature, 32 to 40 degrees, is considered ideal, but leeks seem to tolerate a variety of conditions.

The fanlike leaves of the leek stay green through repeated frosts.

**Onions** Your best keepers will be late-maturing onions with thin necks. Often these are strong-tasting when raw, but they cook up well. You may, like us, be anxious to plant another crop in the onion row, but it's important to let the onions mature in place. When more than half the tops have bent over naturally, you can drag a rake, teeth up, over the row to knock down the last of the tops. European gardeners often roll a barrel down the row to bend over the onion leaves. Then let the plants dry for another week in the ground. Pull them on a dry day and spread them on screens to cure in the sun for three to seven days. Cut off the tops, leaving a one-inch stub and spread the bulbs out in a single layer to continue curing in a dry, shady place for another two to three weeks. Then gently put them in mesh bags, slatted crates, or cartons with holes punched in the sides, and keep them cool and dry (45 to 55 degrees F, 50 to 60 percent humidity).

We visited some gardeners who braid their onions and hang them from rafters. This is a most attractive way to keep onions, and it saves space too, but onions to be braided should be very well cured and dry at the neck or they may rot. Thick-necked onions, which some people call scullions, are generally poor keepers and should be used soon after harvest, not stored. It's a good idea to braid a length of string in with the onion braid for added strength. One imaginative gardener stores onions in worn-out panty hose. She ties a string between each onion and hangs the stuffed stockings in a cool, dry spot. To remove the onions, she snips off the stocking toe and works up from the bottom of the stocking, removing onions as she needs them.

Spreading onions on a screen to cure in the sun.

Onions can stand fairly low temperature and may even be used if frozen, but should not be allowed to thaw and refreeze. If they're quite cold when you cut them, you'll cry less! If some onions do sprout before spring, plant them in a flowerpot and snip the green tops into small bits to garnish other vegetables.

**Parsley, Green Leafy** This is another good wintering-over crop, but one that needs some light, like chives. Parsley plants grow deep taproots, so the best plan is to dig up a young plant at the end of the summer and pot it in your best potting soil. Keep it outdoors until two weeks before you'll start heating your house, then bring it in and put it on a sunny windowsill so that it can become accustomed to indoor conditions before the air becomes too hot and dry. It's satisfying to see how much of your garden you can keep alive, in one way or another, to tide you over the winter. I also sow parsley in a cold frame in spring or early summer in order to have it fresh for the picking well into the winter and early the following spring.

**Parsley, Hamburg-Rooted** This variety of parsley grows long slim roots like carrots. It is especially good in soups. Treat it just like carrots in storage.

**Parsnips** Parsnips are perhaps the hardiest of all root vegetables. That row of parsnips you hovered over last spring, as you watched for the first sign of life from those slow-germinating flaky seeds, has built up sufficient momentum by fall to carry you through till the next planting time. We enjoy parsnips in four stages. First, after several moderate fall frosts, we start digging the roots for hearty fall dinners. Frost improves the flavor tremendously by changing some of the starch to sugar. Always *dig* parsnips. If you try to pull them you may lose half the root. A 15-inch-long parsnip is not unusual. If it is important to you to get as much as possible of each root, you might want to try digging a trench or running the rototiller right next to the row, then digging and pulling the roots over into the soft soil.

Then later in the fall, we spread mulch over part of the row to hold off the cold so the ground will stay open for digging a few weeks longer. Our third stage of parsnip harvest comes in early December when, before the earth closes over firmly, we dig up another section of the parsnip row to keep in the cool cellar over winter. We trim the leaves from these roots and brush off the clinging dirt and, as with other root

You have to *dig* parsnips—you can't pull them up because the roots are too long.

cellar vegetables, feast at dinner on the roots the shovel nicked. These might spoil in storage. Parsnips keep for two or three weeks without covering, but for longer storage we layer them in damp sawdust as we do carrots. Leaves, moss, or sand will work well too. Ideal conditions for parsnips are 32 to 35 degrees F and 90 to 95 percent humidity.

The fourth phase of parsnip harvest comes in the spring when we dig the frosty sweet roots from the just-softened earth. Spring parsnips are good until the plants send out new leaves. A few small leaves around the crown of the roots are nothing to worry about, but when the stems lengthen and the leaves are full blown, the root becomes tough. Just before this happens, you can dig the remainder of the crop and keep it in your cold storage spot to extend the season another few weeks.

**Peppers** Leave space for a basket of sweet peppers when you plan your root cellar. Peppers must be picked before frost. Leave the stems on. We had been in the habit of leaving our basketsful of hastily rescued peppers on the porch for a while, reasoning that cool night temperatures would keep them in good condition. They seem to keep even better, though, at the more constant moderately low temperature of our old dirt-floored cellar, which is close to the ideal 45 to 55 degrees F and 80 to 90 percent humidity recommended for pepper storage. Peppers shouldn't be chilled; a temperature as low as 40 degrees F will encourage decay. Peppers don't keep long in the refrigerator, because most refrigerators maintain a temperature around 40 degrees F. If humidity in your storage area is lower than ideal, you can enclose the peppers in perforated plastic bags. Use the red ripe ones first; they're the most perishable.

Hot pepper plants may be pulled whole and hung in a warm dry place, not in a damp or cool cellar. Around 50 degrees F and 60 to 65 percent humidity is about right for your dry chili peppers. They should last about six months.

**Potatoes, Sweet** Well-stored sweets will keep till spring, but they'll spoil readily if they're bruised or if conditions aren't right for them. Dig your storage crop as late as possible in the fall, because the roots increase in size most dramatically during September and October. Be sure to get the crop in as soon as frost has killed the vines, though. Although we've heard that the sweet potatoes will be off-flavored and will keep poorly if not dug the day after the vines die, we have not proven to our own satisfaction that this is true. Cold nights *are* likely to cause chilling injury to the sweets, though, so we always play safe and harvest ours promptly after the rubbery vines turn limp and black from frost. If just the leaf tips are frost-nipped, we leave the crop in the ground until a harder frost.

Moon-sign gardeners often dig sweet potatoes during the dark of the moon before the new moon. Gardeners in the South, who may want to dig storage sweet potatoes before their late frost, might find it useful to know that the raw inner surfaces of a mature sweet potato, when broken in half, dry quickly, but in one that is still immature they stay moist for some time.

Wrapping sweet potatoes in newspaper after curing.

Begin by digging sweets carefully — ideally from the side of the row — to avoid nicking and slicing too many of them. Then let the roots air-dry for half a day so that you can easily brush the extra soil from them. (Don't scrub or rub, though. Those skins are extra tender.) Always handle sweet potatoes gently, as though they had feelings.

Curing is the next step, and it's especially important for sweet potatoes. Exposing the sweets to warm temperatures (80 to 85 degrees F) helps to toughen the skin, changes some of the starch to sugar, and encourages "corking over" of small skin cuts. Although humidity of 90 percent is recommended for the 10- to 14-day curing period, we've had excellent results simply spreading the sweets behind our wood stove and covering them with a damp (well wrung-out) towel.

To keep the potatoes, we wrap them individually in newspaper and pile them in baskets. Once they're packed away, we try not to handle them until we're ready to use them. All bruised or cut sweet potatoes are set aside for immediate eating. We eat the tiny ones right away too.

They're delicious and tender but they don't keep well. The baskets stay all winter in an unheated room where temperatures range around 50 to 60 degrees F during the day, although it is often five degrees cooler there at night. It's extremely important to keep stored sweets dry, as you may have learned the hard way if, like us, you've tried to store them in the refrigerator. Eighty to eight-five percent is the officially recommended humidity, but our storage room is somewhat drier than that.

When we first started growing sweet potatoes, we had trouble keeping them through the winter until we learned how, so we've made a point of asking other gardeners how they treat their sweets. The following storage arrangements have been used successfully by experienced gardeners.

1. Pack potatoes one inch apart in cartons on a two-inch bed of sawdust and pour sawdust between the potatoes. Do not dampen the sawdust. Spread an inch of sawdust over each layer of potatoes and another several inches on top. Don't let the sweets touch the side of the carton; there should be a margin of sawdust all around.

2. You can bury sweet potatoes in bins or cans of feed grain — oats or wheat will work well — and keep the cans in an unheated room. When you've used all the sweets, you can still feed the grain to the animals. We wouldn't bury sweets or other roots in grain intended for human use because it's hard to be sure you've removed all dirt and

## Outdoor Storage Bin

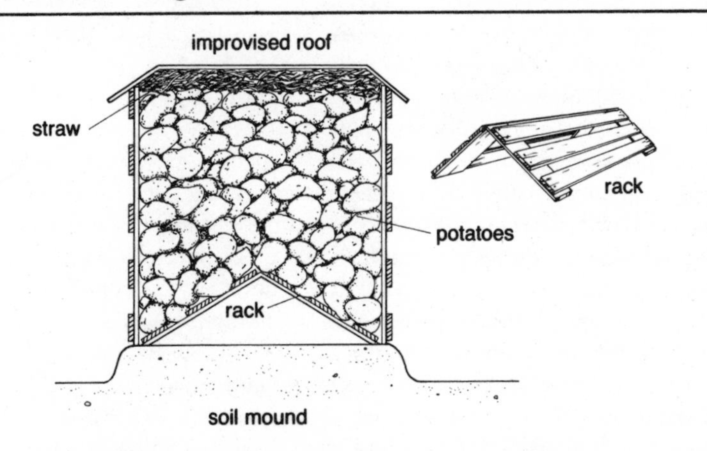

grit from the roots. (Make that whole grain, not crushed, crimped, or cracked. It'll retain nutrients longer if less surface is exposed to the air.)

3. A Midwestern gardener wraps his sweet potatoes in newspaper, packs them in large brown grocery bags and, after tying string around the bags, suspends them from the joists in his basement.
4. Some gardeners who live in dry climates with mild winters construct outdoor storage bins. Here's one version: Prepare a "platform" by firming the soil into a flat-topped mound several inches higher than the surrounding soil. Then place a slat-sided box on the mound. An inverted "V" made of slats nailed to two short boards should be put in the center of the box. Then carefully fill the bin with sweet potatoes, which will rest on the air-admitting rack. Finally, cover the hideaway with dry vines or straw or put a metal roof over it.

**Potatoes, White** Second-crop or late-maturing potatoes are the best for storage. The problem with summer-dug potatoes is that you can't cool them easily, and they shrivel and sprout much sooner in the warm weather. Early potatoes will keep about four to six weeks at a temperature of 60 degrees F. Potatoes are ready to dig when the tops dry up, but they may safely be left in the ground until cool weather, as long as six weeks after the tops have died, if drainage is good. In a very wet fall, though, they might resprout or rot, so keep an eye on them. If some tubers are partly exposed, toss some hay or soil over them to exclude sunlight which causes greening. Parts of potatoes that have turned green contain toxic solanine, so it is important to keep all potatoes covered. If you must dig when weather is warm, do the job early in the morning while the soil and the tubers are still cool.

Potatoes should be cured before storing to give them a chance to heal surface nicks and toughen their skins. Spread them out in a protected place where the temperature is 60 to 75 degrees F. They should not be exposed to rain, sun, or wind during curing. After a one-week to two-week curing period, potatoes are ready for storage. Their skins will have thickened and minor wounds should be scarred over. Potatoes with many harvest injuries produce much more heat in storage and are harder to cool, especially if they have dirt clinging to the skins. Keep them in shallow layers.

For winter keeping, put your potatoes in a cold, damp spot. Unlike cabbage and onions, they're useless for food if they are frozen. Frosted potatoes may be cooked and fed to animals, though. Potatoes keep best at 36 to 40 degrees F with high humidity, around 90 percent. At low temperatures, 35 degrees and below, some of the starch in the potato turns to sugar, giving the spuds a puzzling and not very appealing sweet flavor. This is easily corrected, though. Just bring small batches of the affected potatoes into a 70-degree-F room, and in a week or two the sugar will revert to starch.

Under good storage conditions, which are not hard to provide, potatoes will keep for four to six months. Keep them dark. Light, as well as warmth, promotes sprouting and will also turn the potatoes green. Towards spring, many growers try to keep their potatoes in the lower temperature range to hold back sprouting as long as possible.

Cold potatoes bruise easily, so they should be handled gently when put into storage. It's important too to pile the potatoes in several small mounds rather than in one big heap. When massed in a large pile the bottom potatoes may be bruised by the weight of those on top. In addition, large mounds of produce tend to heat when ventilation can't reach the center of the pile, and that heat lowers quality and shortens storage life. Air should be able to circulate through the pile to prevent condensation. Lugs or bins of potatoes may be safely stacked as long as air can circulate between them. Humidity should be high. Although potatoes lose some moisture through respiration, the main cause of shrivelling in storage is low humidity. (See chapter 14.) It's a good idea to cover piles of potatoes with sacks, straw, or shavings to prevent condensation, which can invite spoilage.

Most experts recommend keeping stored apples and potatoes separate, because apples give off ethylene gas which promotes ripening, maturation, and — in the case of potatoes — sprouting. In a small-scale home setup, though, it is not always possible or practical to segregate the spuds from the apples. If you use a decentralized system (see chapter 12), you can usually manage this, but if you have a single root cellar, you probably keep everything there. Having read so many warnings about the danger of keeping potatoes and apples together, we've been making a point of asking root cellar owners what their experience has been. A majority of those we visited and consulted *do* keep apples and potatoes in the same room. For most of these people, sprouting of potatoes has not been a problem.

Why aren't these potatoes sprouting prematurely as they "should"? (Most potatoes start sprouting by spring.) We think that ventilation is the answer. In a well-ventilated root cellar (see chapter 13), the ethylene gas fumes will be carried off by air currents rather than settling down around the potatoes. But if you were to shut a bag of apples in a

Styrofoam cooler with a bag of potatoes, you'd have sprouted potatoes a lot sooner. So if, like many people, you're counting on apples and potatoes as the mainstay of your root cellar larder, I wouldn't worry about it. Arrange a well-ventilated cellar, don't pile the apples and potatoes right next to each other, and keep the potatoes covered with newspaper or burlap bags if you wish. This we know: There are plenty of root cellars from Maine to Michigan to Oregon where potatoes and apples get along just fine together.

**Pumpkins** Treat pumpkins like squash. Technically, pumpkins do well with more humidity than squash — 70 to 75 percent rather than 60 to 70 percent — because their skins are slightly more tender. Also for this reason, they don't last quite as long in storage as do squash. High storage temperatures will make them stringy. Pumpkins that have lost their stems won't keep well. Cook and use them in the fall or cut them in thin slices and dry them.

Always leave the stem on the pumpkin when harvesting. Beautifully put together, isn't it?

**Radishes** Winter radishes are the ones to keep; summer radishes are perishable. These large hearty good keepers are a bit more fibrous than their quick-growing summer cousins, but they'll last until February if well stored. Trim off the leafy tops close to the point where they join the root. Store them like carrots — either well mulched in the ground (be sure to cover their protruding shoulders well), or layered in moss, sawdust, or sand in your coldest above-freezing storage place, along with carrots, beets, potatoes, and other root crops. They'll keep for a week or so without covering but will shrivel if left unprotected by some damp covering for much longer than that.

**Rutabagas** The Swede turnip, as these large sweet roots are also called, will last two to four months in storage. They shrivel more readily than carrots, and so must be kept moist. Burying them in damp sawdust, sand, or moss seems to work best. You could also wax them to retard moisture loss, as stores do, but we'd use beeswax for this, not paraffin or other hydrocarbons.

**Salsify** We treat salsify much like parsnips, waiting impatiently for the first fall frosts to nip the plants and improve the flavor, then eating part of the row fresh, later mulching to keep the ground open, then dashing out in November or December to dig up a basketful of roots just before the ground freezes hard. Salsify is also good in spring before new growth starts. You can dig the rest of the row then and keep it in the root cellar for another few weeks. The slimmer roots shrivel more readily in storage than parsnips, carrots, or winter radishes, so use them rather soon after bringing them in. Surrounding them with damp sawdust, sand, or moss helps. Like other root crops, salsify likes to be just above freezing (32 to 40 degrees F) and very moist (90 to 95 percent humidity).

**Scorzonera** This black-skinned, white-fleshed look-alike is not closely related to salsify, but for winter keeping should be treated the same way.

**Soybeans** Green soybeans are usually ready for picking in September. If you leave them in the pod, you can keep them in a cool, dry place (35 to 40 degrees F, 60 to 70 percent humidity) for two to four weeks and steam and pop them for the table as time permits. Dry soybeans should be treated like seeds and grains. (See chapter 11.)

**Squash** Well-stored winter squash will keep till spring and need very little attention. Start by picking them at the right time — when they're mature the skin will be so hard that your fingernail can't puncture it. Always leave stems on, as the soft scar on a stemless squash is likely to invite spoilage. Painting the scar with beeswax helps to prolong the storage life of stemless squash, but they should still be used promptly while they are sound. If some squash are immature when picked, they should also be used promptly, for they will be the first to shrivel and collapse. If squash borers kill the vines in early fall, the growing fruits may not mature well. Use them up early, too. Be sure to gather your squash harvest before heavy frost strikes.

After picking comes curing. During the curing process the squash loses some of the moisture in its flesh and develops a hard rind, both aids to longevity in storage. Note: The acorn squash is an exception. They like lower temperatures than other squash and should *not* be cured. Keep them at 45 to 50 degrees F, not over 55 degrees, or they'll turn stringy and dry. Acorn shells should stay green. If they turn orange, they've lost quality, probably from high storage temperatures. They don't keep as long as other winter squash varieties.

To further harden the rind, spread the squash on the grass or on an open porch where the sun can reach them. They should be covered if frost threatens or brought in if heavy frost is likely. If weather is rainy, your squash would probably cure more effectively indoors at 70 to 80 degrees F. A spot near a box stove or furnace or hot water heater might be just right. Squash should cure for 10 to 14 days before they are ready to store. Handle them gently and keep them warm and dry. Fifty-five to sixty degrees F and 60 to 70 percent humidity is ideal. At higher temperatures the flesh becomes stringy. Avoid heaping squash in piles or they may heat and bruise each other. It is best to keep them spread out in a single layer, an inch or so apart. Some gardeners keep a resilient layer of hay or dry leaves under their stored squash, and slatted shelves allow good air circulation. Good root cellars are too damp and cold for squash and pumpkins, but there are other places to keep them. We find an unheated side room just right. Attics, regular heated basements (far from the furnace), and spare bedrooms are often within the ideal temperature range. Some folks stash them under the bed. Ruth Stout once wrote that she kept hers in a box under her kitchen table. Good storage squash will keep for as long as six months — a lot of good eating from an undemanding vegetable.

**Swiss Chard** Chard has its own place in the year-round gardening scheme. While it isn't a storage vegetable as such, it can be kept bearing into the winter and so earns a place in any year-round vegetable scheme. A good way to handle spring-planted chard is to cut off all leaves in August, leaving a two-inch stub. Tender new leaves will grow in the fall. Light frosts don't affect the chard. We know a clever gardener who keeps her chard going well into December by protecting part of the row with a tent of clear plastic supported by a framework of heavy wire semicircles poked into the ground. Chard has such a deep, strong root that it lives over winter and is often one of the first garden greens to appear in the spring. Where winters are mild, chard will remain green and ready to pick all winter. Greens are so nutritious that any plant that can extend the fresh-greens picking season is worth our attention.

**Tomatoes** Frost needn't end your tomato season. Mature green tomatoes may be ripened indoors. The best tomatoes for ripening inside are those from youngish plants in their prime, rather than plants that have been bearing all season. For us, late-starting volunteer tomato plants often serve this purpose. Some gardeners start a special early summer planting of a few tomato plants so they'll have vigorous vines to provide late-fall tomatoes. Or you can root cuttings from established plants in the summer. The slight extra trouble is worthwhile when you can slice your own red, ripe tomatoes into a salad on Christmas day. Leave the small, fluted, white tomatoes on the bush. For storage, pick only those that are mature green or riper. (A mature green tomato is well developed, shiny, and medium or deep green. As it ripens, it gradually turns pink and then red, or yellow and then orange in the case of Jubilee, Sunray, and other orange varieties.)

Try to pick the tomatoes without stems, or remove the stems after picking so that they won't puncture other fruits. Pick the fruit before frost, because chilling injury can prevent further ripening in green tomatoes. Riper tomatoes are less sensitive to low temperatures. When you bring the tomatoes indoors, sort them and keep the ripest ones separate so they won't be bruised by the harder green fruits. Hold out some tomatoes to ripen quickly at room temperature. Spread these in a single layer out of the direct light in a room where the temperature is 60 to 70 degrees F. Mature green tomatoes will ripen in about two weeks at room temperature. Those that have started to turn red should be ready within a week. You can wrap the tomatoes individually in paper to protect them and keep out the light. Don't scrub or rub them to remove dirt, as they're easily bruised. According to the United States Department of Agriculture, it's all right to wash them gently if you dry them before storing, but we've never found this necessary.

French gardeners often ripen tomatoes on beds of straw in the cellar. Ripening tomatoes will do better if the air is not excessively moist. Check the ripening fruits every few days to catch any that are going bad and use the good ones at their best.

If you have a lot of green tomatoes, you'll want to store some to ripen gradually later. Try not to chill these storage tomatoes below 50 degrees F for more than a day or so or they may take a chill and never ripen. Fifty-five to sixty degrees will keep the fruits on "hold" for several weeks. Mature green tomatoes will ripen in 25 to 28 days at 55 degrees, or you can speed up the process and bring a few fruits at a time into a warmer 60- to 70-degree room for gradual ripening.

Ethylene gas promotes ripening and is sometimes used commercially to ripen mature green tomatoes. Tomatoes release ethylene gas as they ripen, but it is sometimes possible to further encourage the ripening process by placing a few apples — which also emit ethylene gas — among the green tomatoes. This is particularly effective in a closed-in space with little air circulation to carry the fumes away.

An even easier way to bring tomatoes indoors for ripening is to pull up the vines and hang them in a basement or shed. Some nutrients probably remain in the leaves and stems to continue to nourish the fruit for a while after picking, and because the tomatoes are well ventilated and free from pressure they ripen well when hung. The vines and fruit should be dry when pulled, or rot may set in. You can hang the vines on nails in the rafters, tie them up, or drape them over a rack standing on the floor.

Either way, you can often coax another four to six weeks of fresh eating from the tomato patch after frost. You might also want to try to keep an especially nice tomato plant producing right there in the garden for several weeks after frost by surrounding the plant with hay bales and placing a storm window over the top of the hay bale enclosure. This is a good trick to try with the variety Sweet 100, which produces delightfully sweet little cherry tomatoes that are just great for school lunches.

Plant breeders are working on developing tomato plants that are more resistant to cold and other kinds that produce fruit that keeps unusually well, so watch the seed catalogues. Someday soon it may be possible to grow good storage tomatoes!

**Turnips** You can leave turnips out in the frosty garden row later than rutabagas, but not as late as carrots unless your winters are fairly mild. Bring them in before a heavy freeze, lop off the tops, and then treat them just like carrots, storing them in damp sawdust, sand, or moss. They also keep well in buried boxes and barrels, and in leaves in the barn or garage. Your storage turnips should be grown from a summer planting, for if they are held over from the spring crop, they will be strong flavored and woody by fall. Incidentally, it was the turnip, not the pumpkin, that was probably the first jack-o'-lantern. In Scotland the custom of putting a lighted candle in a hollowed-out turnip or rutabaga evolved into the Halloween tradition of carving a face in the side of a scooped out vegetable and ultimately into the widespread use of the pumpkin, which was plentiful here in the New World. (And somewhat easier to carve, we should think!)

# 9

# Fruits

Picking pears is a pleasure that has to be learned, like many other pleasures.

Count de Comminges
*Laura's Garden*

Even a single backyard apple tree can yield several bushels of fruit for storage, and if you add grapes and vine fruits from the garden, pears from a neighbor, perhaps some pawpaws or crabapples from the hedgerow, and citrus fruits purchased by the case, why, you've got a fruit cellar too!

Some people make an effort to keep their fruit separate from their vegetables. Fruit has a delicate flavor that is sometimes affected by strong-tasting cabbage family vegetables. Moreover, after picking, most fruits give off ethylene gas that tends to promote sprouting in potatoes and shorten storage life of some other vegetables. Most of the practical gardeners we consulted keep their fruits and vegetables together, though, and we also intend to do this in our new root cellar. It's easier to make other arrangements for your cabbage and keep your fruits and carrots, beets, potatoes, and so on in one cellar, unless you use decentralized storage setups. Fruit makes a simple, wholesome winter dessert, and it's well worth the trouble to keep it properly. Wrapping individual fruits in paper helps to cut down on the absorption of alien flavors.

**Apples** The apple is the queen of storage fruits. We know people who have built root cellars just to store apples and potatoes. Apples that have matured on the tree are your best bet for storage. When harvested green, they tend to shrivel in storage and seem to be more likely to be affected by diseases like scald and bitter pit. Summer apples are relatively fragile, and they should be canned, frozen, or dried for long-term keeping.

Leave the stems on the apples. Pulling them off could break the skin and invite spoilage. If you pick your own apples, you might want to be aware of a fine point that British gardeners often take seriously. The "king" apple, the center one in a group of fruits, is usually the largest because it draws directly on the sap. Sometimes these large central apples have thick stems. They often seem to have a shorter storage life than the other smaller apples, which is a good excuse to munch on one as you're picking! Some gardeners remove the king apple by hand thinning before the fruit develops fully.

Early fall apples like the Cortland and McIntosh and the mid-fall Grimes Golden keep well for a month or two if you can keep them cold enough. The trick is to chill your storage area sufficiently in September and October to hold these early pickings in good shape. It helps to leave your apples out overnight in the cool evening air and then stow them away in the early morning while they're still cool. For over-the-winter storage, count on late-ripening apples that are harvested during cooler mid- and late-fall weather when your root cellar will be cold enough to maintain their good condition. Folklore has it that apples are less likely to rot or bruise in storage if picked in the dark of the moon.

Apples keep best at 32 degrees F with a relative humidity of 80 to 90 percent. Near 32 degrees, a humidity reading of 80 to 85 percent is enough to prevent shrivelling, but at higher temperatures, 40 to 50 degrees, the humidity should be closer to 90 percent. Higher temperatures shorten storage life considerably. At 40 degrees F, for example, apples ripen twice as fast as those held at 30 degrees. At 50 degrees, the aging rate doubles again, and apples held at 70 degrees deteriorate twice as fast as those kept at 50 degrees. If you can keep your storage area cold, you should have apples until spring. If your root cellar temperature remains in the low 40s, you can still store apples — they just won't keep as long. Use them and enjoy them in fall and early winter.

When they become softer than you like for fresh eating, serve them as baked apples.

The experts will tell you that you can't store windfall apples. Well you can, for we've done it. You just need to use a little common sense. Weed out the soft or badly blemished ones before putting them into storage, and keep them in small lots — one peck or so in a container — so you can check them frequently. Use them, don't just admire them! They won't last until spring, but we've kept good early windfalls until well after Thanksgiving. They should not be mixed in with your other apples, of course. If you don't grow your own apples, and you value unsprayed fruit, windfalls can be a real find because the trees from which they fell are often old backyard apple trees that no one has bothered to spray. If you like to trade, as we do, you can often work out some interesting deals to get some of these good, often nameless, apples. We've traded squash, scrapple, goat milk, and our own labor (cleaning up dropped apples under the tree for the owner who had a bad back) for some great windfalls. Somehow they tasted better, too, because they were unsprayed, and the transaction was a personal one.

Apples just naturally soften as they continue to ripen in storage. Sometimes they also have other problems. Scald is one of the most common storage disorders in apples. Scalded apples have a browned or burned discoloration on the skin, often appearing first on the green side of the apple. The flesh underneath the scalded spot turns mushy. Apple specialists don't know much more about apple scald today than they did 55 years ago, but scientists still seem to agree that the condition is somehow caused by gases given off by the apple itself. Ventilation should help to control scald. Overripe apples sometimes succumb to blue mold, brown rot, and other microbial invasions. Just weed them out and spread out the good ones. Apples that have been bruised are susceptible to internal breakdown. Shrivelling is caused by dryness in the storage area, and/or by too high temperature that sets the apple to using up its reserves through rapid respiration. An apple that's 85 percent water will shrivel when it's lost 5 percent of its moisture content, so you can see that the line between soundness and wrinkling is rather fine. Shrivelled apples are all right for cooking, but less desirable for eating fresh, and they tend to spoil sooner than unshrivelled fruit.

Apples should be kept in fairly shallow layers, as bruising can be a problem in large piles of fruit. Half-bushel baskets or slatted crates work well.

Commonly available apple varieties that keep well into the winter include Baldwin, Winesap, York Imperial, Jonagold, Jonathan, Northern Spy, Rome Beauty, Rhode Island Greening, and Arkansas Black. McIntosh, Cortland, and Grimes Golden will keep for several months under good conditions. Many heirloom apple varieties are especially good keepers. These old favorites aren't grown for commercial production of fruit any more, but they may be ordered from nursery catalogues. (See Sources.) The following is a partial list of old apples that have a long storage life:

- American Beauty
- Ben Davis
- Lady Apple
- Rusty Sweet
- Sheepnose
- Sutton
- Winter Rambo
- Pound Apple
- Winter Smokehouse
- Yellow Bellflower
- Golden Russet
- Spitzenburg
- Winter Banana
- Roxbury Russet

In the old days, apples were often wrapped in paper, then packed in barrels that were covered with straw and kept in a warm part of the barn.

**Cantalopes** Short-term storage only is possible for this fruit, but they're good while they last. Let the melons ripen on the vine. They gain little if any sweetness after picking if harvested green. Ripe melons have prominent netting, light yellowish tan ground color, and a pleasant aroma. The stem should separate readily from the melon when pressed with the thumb. Melons for cold storage should be picked in the morning before the sun heats them up. Pad the garden cart with hay or rags to prevent bruising. Ripe cantalopes will keep for 10 to 14 days at 35 to 40 degrees F and around 85 to 90 percent humidity. Nearly ripe 'lopes that must be gathered when frost threatens may be buried in a bin of oats or other grain in a cool outbuilding. These will last four to six weeks. They'll never be as sweet as the vine-ripened melons, but they make an excellent addition to a mixed fruit cup.

**Crabapples** Crabapples keep for several months if picked from the tree. They may be hung in net bags in the root cellar. They require the same temperature and humidity as apples.

**Cranberries** Should you have occasion to store more than the usual one-pound packages of cranberries available in markets, keep them cool and moist but somewhat warmer than apples — 36 to 40 degrees F.

**Dried Fruit** Keep in a cool *dry* place, well sealed in jars or cans.

**Grapefruit** With its thick skin, grapefruit keeps well for a month or even two months in a cold damp place — 32 to 40 degrees F and 80 to 90 percent humidity.

**Grapes** Fall-ripening grapes will keep in the root cellar for a month or two. Catawba, Sheridan, and Steuben are known as especially good keepers, but if you have other varieties, don't hesitate to try them in storage. The grapes should ripen on the vine. Hold them by the stems when you pick them to avoid bruising the fruits. We cut grapes from the vine with scissors, because when they are ripe, they often fall off if you try to pull them from the vine by the stem. According to horticulturist Dr. George Abraham, grapes are especially sensitive to chemical weed killers and may ripen unevenly if affected by drifting sprays. Grapes need sun to ripen and if the vine is unpruned, they may ripen late because there are so many of them.

Keep grapes cold and fairly moist — about 40 degrees F and 80 percent humidity. They should be spread in shallow layers, one bunch deep, in trays or baskets. Better yet, follow great-grandfather's practice of suspending individual bunches from dowel rods held in place by cleats attached to the underside of a shelf or wooden box top. This spreads out the grapes and helps to prevent mold by encouraging good air circulation. An old grape storage trick used by French gardeners involves suspending the individual bunches from hoops hung from the ceiling — kind of a grape mobile.

**Melon Squash** This unusual fruit looks like a long-necked squash but tastes somewhat like a melon. The flesh is crisp and firm. It is well worth growing for a winter fruit supply, especially if you enjoy desserts of mixed fruit. We like it sliced into a fruit cup, where its crisp texture and orange color perk up our home-canned pears and peaches. The melon squash should be left on the vine until frost threatens because they need a long ripening season. Here in south-central Pennsylvania they are still green at picking time in October, but gradually ripen in storage. The rind turns orange-buff, and the melon is actually sweeter after two or three months in storage than it was when picked. The fruits are large, but we've found it possible to cut off a few inches to serve, cover the cut surface with an elasticized plastic storage-dish cover, and keep the cut melon in a cold place for several weeks as we continue to cut from it. A temperature of 50 to 60 degrees F in fall and 40 to 50 degrees in winter

## Grape Storage

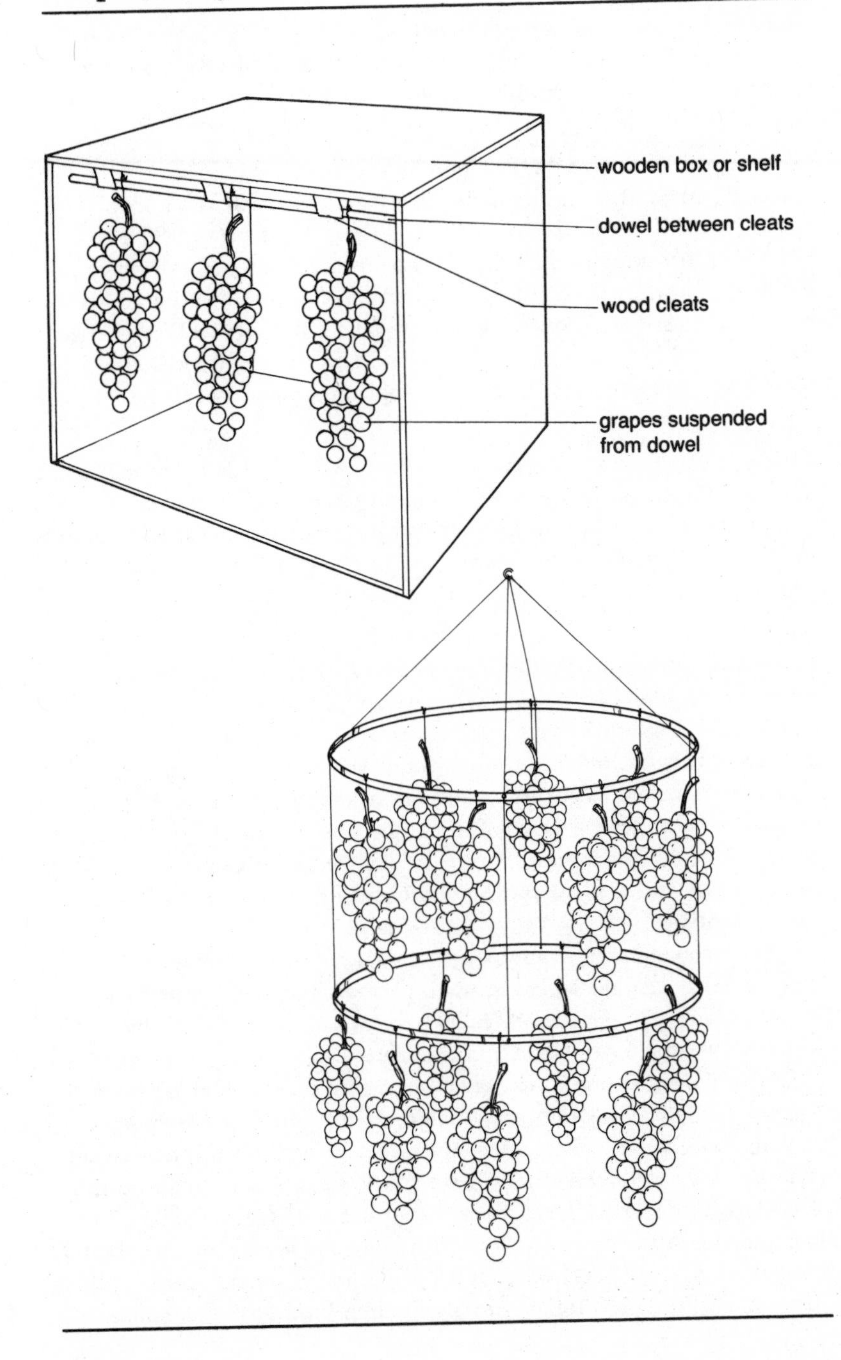

seems to suit the melon squash. That, at least, is the uncontrolled temperature of the room in which we have kept it. Probably a somewhat warmer temperature would be better in winter.

**Oranges and Tangerines** Often purchased in bulk at Christmas time, or especially ordered from organic suppliers, oranges will last a month or two in a cold damp place — 32 to 40 degrees F, 80 to 90 percent humidity. Browning of the skin in oranges sometimes occurs when they are held at or near 32 degrees. Sort the fruits from time to time to weed out any that are squashy or affected by blue mold. We've kept oranges for two months in a cold pantrylike room far from the wood stove. The room was not especially damp so we kept the oranges covered.

**Pawpaws** Pawpaw trees grow wild, often in semishaded thickets, and the fruit they bear is sweet and smooth-textured. Pawpaw fruits are sometimes called custard apples. If you pick them from the tree before they are dead ripe, you can store them in a cold place — around 35 to 40 degrees F, for several weeks.

**Pears** If you have a root cellar, then you won't have to can all your extra pears. Pears should be mature when picked, but they should not ripen on the tree, or they will have a gritty texture. Sometimes they also rot at the core when left on the tree too long. When *should* you pick pears? When the skin has begun to change from definite green to a suffused yellow-green, and the fruit separates easily from the tree when you lift it gently. Start picking from the sunny side of the pear tree first. Leave the stems on the fruits, but handle them carefully so they don't puncture their neighbors.

It's a good idea to wrap storage pears individually in paper. Spread them in fairly shallow layers in small cartons, and keep them cold and moist — 32 to 40 degrees F, 80 to 90 percent humidity. Then, when you want to eat some, bring them into a warmer room and ripen them at 60 to 65 degrees for several days. If kept too warm during ripening (over 75 degrees), though, they may spoil before they ripen. Good storage varieties include Bartlett, Bosc, Anjou, Devoe, Winter Nelis, and Kieffer. (Kieffer, however, is coarse and best for cooking, not out-of-hand eating. The Red Bartlett pear has a thicker skin than the regular yellow Bartlett and one fruit grower we talked to thought it was an even better keeper.) Don't try to store Clapp's Favorite; it softens quickly after harvest. Well-treated storage pears will often keep until Christmas.

**Plums** Japanese plums may be held at about 40 degrees F in a damp place. Late-ripening plums usually keep best, because root cellars are not always cold enough for keeping summer plums. The Italian prune plum Fellenberg also keeps well. Put plums away in shallow layers to prevent bruising.

**Quince** All quinces keep well. These hard, knobby, fuzzy-skinned cousins to the apple aren't sweet or juicy enough for fresh eating, but they have a melting, delicious flavor when cooked or made into jelly. Let the fruits ripen on the bush or tree until they've turned yellow. When ripe, they have a very pleasing aroma. In fact, quinces were once tucked into clothes closets and drawers as sachets.

Store quinces as you would apples, at 32 to 40 degrees F and 80 to 90 percent humidity. They'll usually last until spring.

(Did you know that you can often persuade a flowering quince bush to bear fruit if you put a bouquet of quince blossoms from a fruiting quince tree near the bush while it is in flower? All it needs is a little help with pollination, and the bees do that. The resulting fruits are good to eat, cooked as you would cook any other quince.)

**Watermelon** Harvest the fruits when they're good and ripe. Thumping them isn't always conclusive, but it helps, along with other signs. An unripe watermelon sounds tinny or metallic. A ripe one makes a deeper plunking noise when you knock on its side. Also look for a dry tendril, a white or yellow patch on the underbelly where the melon touched the ground, and a surface bloom that replaces the skin's former gloss with a duller sheen. Watermelons have such a thick rind that they'll last for two or three weeks. And if you're determined to serve watermelon for Christmas, grow the Winter Melon — an unusually good keeper.

# 10

# The Underground Garden

Roots: the prime decision-makers for life on this planet.

Fra de Berlanga
quoted by Charles Morrow Wilson
in *Roots: Miracles Below*

In addition to the solid, hearty mainstay roots — the beets, carrots, and potatoes — of any root cellaring program, there are certain kinds of roots that can provide you with tender, delicate, brand-new green salad shoots. These fresh-cut morsels make super first-hand eating during winter's coldest months. They are, in fact, gourmet fare that can only be found — and then irregularly — in the most expensive city grocery shops. But you can grow them yourself, from homely roots you've set aside for this purpose.

We're talking about the process of forcing roots, which we prefer to think of as encouraging the roots to do what comes naturally to them — sprout. Fleshy roots store up tremendous reserves of energy, and when they've been frozen to break their dormancy, brought into a warmish place, supported by soil, and watered, they put all that stored goodness into making new green shoots. That's what forcing vegetables is all about. The leaves and stalks you get in this way are pale and tender because they've been grown indoors away from the light. While it is true that deep green leaves are richer in vitamins, these delicacies still contribute a texture and food value that would otherwise be difficult to come by in January. And they are delicious!

**Asparagus** If your asparagus bed needs thinning, here's a way to get a bonus from those extra plants (or you could raise plants from seed especially for forcing). Large roots are best. They should be at least three years old. Dig the roots up in the fall before the ground freezes hard. Replant them in boxes or pails of soil. Keep them in a cold shed or garage for several weeks. Freezing helps to encourage them to break out of their dormancy, but it is not essential. To encourage roots to sprout into short, tender asparagus spears, bring the planted box or bucket into a warmish place, 60 to 65 degrees F. Water the soil about once a week but be sure that it does not get waterlogged. Keep the box warm. Light won't hurt, but it is not necessary. Snap off the spears as soon as they appear. You won't be overrun with them, but a dozen roots should give two people a taste of spring for several winter weeks.

**Belgian Endive (also called Witloof Chicory)** If you've ever shaken your head over the price of this delicacy in some swanky specialty store, you'll find it hard to beleve that these creamy crisp sprouts are so easy to grow. You need to plan ahead, have a supply of old buckets and bureau drawers, and set aside a few square feet of space where you can mess around with soil and sawdust, but other than that it's just a matter of following these simple steps:

1. Plant seeds in the garden in May and June, so you'll have well-developed roots to harvest in the fall. Deeply worked soil will grow good roots, but it needn't be especially rich. You can save garden space by sowing the seed in a wide band. Thin the young plants to stand about four to six inches apart. The roots are what you're after here. The plants have deep green — but very bitter — leaves in summer. The chicory needs no summer care other than weeding or mulching to keep down the competition.

2. Dig up the roots in the fall just before the ground freezes hard. Trim back the leafy tops to one-inch stubs and shorten the roots to a uniform seven to ten inches. If the roots experience several good freezes before you start to force them, they'll sprout more readily. Don't leave

Trim the tops off witloof chicory roots before replanting them in sand in the root cellar.

them bare and exposed once you've dug them, though. Keep them in sand or sawdust in the root cellar until you're ready to start them on their way to salads.

3. Replant the roots in your motley collection of leaky pails, burst barrels, knobless dresser drawers, and split wastebaskets. These household rejects make good chicory planters if they are deep and provide drainage. A completely sound container should probably be punc-

tured on the bottom so excess water can drain off. Your container should be 18 to 20 inches deep. Put a 10-inch layer of damp soft soil in the bottom, then poke about a dozen roots into the soil. Pack them tightly together. Crowding won't hurt them and the quality of the soil you use isn't important. All the power is in the roots. Water the roots thoroughly, spread a 6- to 8-inch layer of damp sawdust over the soil surface, and cover the planter with damp newspaper. You can use sand rather than sawdust to cover the roots but our experience has been that sand is much more difficult to wash off. Sawdust floats away; sand, on the other hand, seems to wash more deeply into the crevices of the sprouts.

You'll need a temperature of at least 50 degrees F to induce sprouting. Aim for 50 to 60 degrees. A first-floor room somewhat out of range of the wood stove seems to be just right for us. Water the roots once a week. They start to sprout, a few at a time, in three to five weeks. When the tips start to show through the sawdust, you'll know that the crisp, delicate, blanched salad heads are ready to eat. Cut them from the root and serve them that very day.

You can even get a second cutting of smaller shoots if you are careful not to damage the root crown when you cut off the first shoots. And, although the roots are too worn out to force again after you've sprouted them once or twice, you could roast and grind them for chicory coffee before you discard them.

If you don't have any deep containers, you can put the roots down flat, side by side, on a three-inch layer of soil and cover them with the damp sawdust, as described above.

For early spring crops from roots left in the garden row, heap manure and straw or sawdust over the row and watch for the pointed shoots to poke through.

**Dandelion** Freebie salads can come from the lawn. Dig up good strong dandelion roots in the fall and treat them just like witloof chicory, proceeding from step two on. You'll get tender blanched sprouts in about the same time.

**Rhubarb** Though a little bulkier, rhubarb is no more difficult than Belgian endive. The important thing to remember is that rhubarb must live through a good freeze before it will sprout. So you dig the roots in late fall while the ground will still admit a shovel. Large, strong roots, two to three years old, are best. Some gardeners refrain from taking spring stalks from roots they intend to force the following fall. Leave a ball of soil clinging to the roots, remove the dead leaves, and put the roots in a bucket or carton. Keep them in a shed or garage where they'll be protected from rain but not from cold. The roots should hibernate in the cold for at least six weeks. Then, when your thoughts are turning toward spring, lug in the lumpy buckets and keep them in a warmish place — 55 to 60 degrees F. Light won't hurt, but it isn't necessary. Within a month the tender pink stalks will push up under crumpled yellow leaves. Strong roots should yield about two pounds each.

You can also force rhubarb in the garden by placing a bottomless bushel basket or crate over each crown and heaping manure around the basket. Cover the open top with glass or clear plastic to keep the heat in.

**Sea Kale** This cabbagelike green vegetable may be persuaded to send up four- to five-inch shoots if treated like witloof chicory. Be sure to keep it in the dark, as the leaves are bitter when exposed to the light.

**Parsnips** These leaves are edible. If you have extra parsnips, or some are starting to sprout anyway, poke the roots into a tub of soil and pick the greens. They needn't be covered with sawdust to blanch them.

**Beets** Treat these like parsnips if you have extra or some are going soft.

# 11

# Other Good Foods to Keep in Natural Cold Storage

Store what you use and use what you store.

Esther Dickey
*Passport to Survival*

Good as they are, roots and fruits and sprouted leaves aren't the only foods you can keep in energy-free cold storage. Add some baskets of nuts, jars of popcorn, cans of sunflower seeds, crocks of sauerkraut and pickles, eggs, a hanging ham, and perhaps even a mushroom log over there in the corner. That's rich living, no matter what the bank account says.

To round out your food cellar and add nutrition and interest to your winter meals, here are more foods you can raise at home and stash away in a corner where conditions are right.

## Nuts

A bushel or two of nuts in the shell is a fine resource to have on hand. Cracking nuts is a relaxing leisure activity for a rainy or snowy

Black walnuts and hickory nuts are kept over the winter on a high shelf in our open log shed.

day. In our family, we all look forward to going nutting. This early fall ritual is never a chore. It is good to pick up nuts promptly before squirrels find them. Recently fallen nuts are still quite moist inside. If you crack one you'll find that the kernels are soft and almost juicy. They're difficult to shell out then, and they taste green. Nuts need to be dried for one to three weeks before they're ready to use or store. Spread them out in a shallow layer on screens or trays and keep them in a dry, well-ventilated spot to dry. Always remove husks promptly, or the nuts will mold.

It's important to remember that nuts contain a lot of natural oil that will turn rancid after long exposure to the air or too-warm storage. For this reason it is best to leave the nuts in the shell until you're ready to use them. They take up more space that way, but they'll last longer. Nuts in the shell should be kept in a cool dry place, ideally 32 to 36 degrees F with a relative humidity of 60 to 70 percent. They should be used within a year. Shelled nuts from fall will last all winter (or for six months) if kept cold, or refrigerated, but they'll go stale before summer. They'll keep for two years in the freezer, though, so you might want to do some steady shelling towards late winter to save part of the crop for summer use.

Mice love nuts. Once we returned from a short trip to find that the mice in our old house had apparently spent the whole time we were gone stuffing the hollow in Mike's mandolin with 92 hickory nuts we'd left out in our storeroom. Moral: keep nuts in a covered can or other mouseproof container.

It's a good idea to punch a tiny vent hole or two in the can in which you keep your nuts to prevent mold. Hickory nuts we've stored in unventilated cans have remained in good condition, but they were very well dried when we sealed them away. Nuts that haven't been dried for two weeks or so after husking are sure to mold when tightly covered.

We keep some unshelled nuts in covered cans in a cold unheated room in the house where they stay in good condition all winter. This storeroom is hot in the summer, though, so we must transfer the nuts to a cooler place in the spring. Several bushels of nuts stay on a high shelf in our open log shed over winter. It doesn't matter if they freeze; the

**Nut Cage**

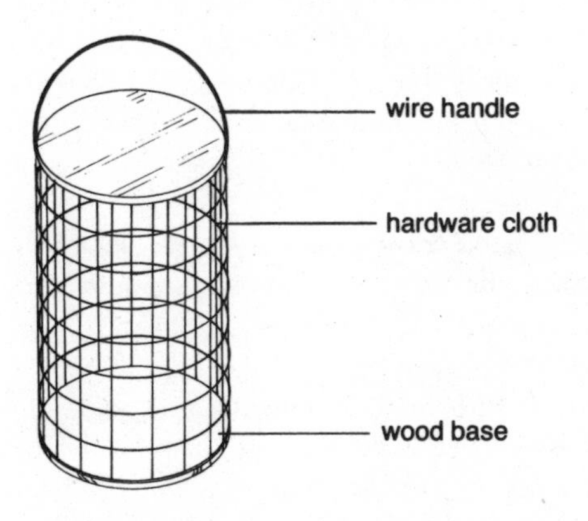

colder the weather, the better they'll keep. Squirrels do get some, but there seem to be plenty left for us.

If squirrels or mice are likely to make serious inroads on your supply of stored nuts, you might want to make up some of these nut cages designed by Mr. Baird Hershey of York Springs, Pennsylvania. Mr. Hershey, an experienced nut grower, fills his storage cages with nuts and hangs them on his screened-in porch, where they're kept cold but not exposed to rain or snow.

Cut a circle of wood 8 inches in diameter for the bottom of the cage. Then cut a rectangle of ½-inch galvanized wire mesh (hardware cloth) 20 inches wide and long enough to go around the perimeter of the wood circle. Staple the wire to the wood base and wire together the sides of the mesh cage. Hershey uses copper wire (so it won't rust) to make a handle at the top for hanging the nut basket. Fit a cover of mesh or sheet metal over the top.

Should you find yourself dealing at any time with a batch of nuts in which many are empty or have shrivelled nutmeats, you can cull out the duds by putting the nuts in water. The empty ones will float.

**Chinese Chestnut** Pick up the nuts regularly as they fall, so they won't mold. We found it necessary to pick daily to keep ahead of our dog, who loves Chinese chestnuts and was eating every one until we discovered what she was up to. Nuts usually fall free of the burrs when they hit the ground. Chinese chestnuts differ from other nuts in having a relatively thin shell and a higher percentage of carbohydrate. Probably for this reason, they need a bit more dampness in storage, and even with dampness they'll last only a month or two, and seldom more than three. So use them and enjoy them early. The best way to store them is to keep them in an open plastic bag or a closed paper bag. Chinese chestnuts shouldn't be air-dried for more than a few days, either, or they'll become impossibly hard.

**Filberts** Dry filberts on trays for a week or so before storing them. Remove all nuts from the papery husk that surrounds them. They usually fall out freely.

**Hickory** Shagbark and shellbark hickory trees yield delicious nuts. We pick them up as they fall and then sit there in the meadow with the sheep and snap off the dry four-part husk so each nut can dry thoroughly. These seem to dry well for us in half-bushel baskets left open to the breeze.

Both hickories and black walnuts have extremely hard shells, but they may be cracked with a hammer. We use a gear-operated nutcracker designed especially for these tough-shelled nuts. It shells out hickories in halves and quarters (and some fragments too, of course) and black walnuts in quarters and large fragments. We wouldn't be without it. We've never seen one in a store, but they may be ordered directly from the man who manufactures them. (See "Sources.")

**Pecan** Late-ripening pecans must sometimes be persuaded to drop by bumping or shaking the branches. Be sure to spread a sheet or tarp under the tree first so the nuts will be easier to find. Dry and store like other nuts.

**Black Walnut** Tons of these good nuts fall on country roadsides and town streets each year, many of them never used. If you don't have a black walnut tree, you might very well find it possible to forage or barter for some of these good nuts. We husk ours promptly so the nuts can begin to dry. The husks contain a long-lasting brown stain — once used as a dye — and they'll stain your hands. In many country communities husk-stained hands are accepted as normal in October. The nuts needn't be husked by hand, though. We spread ours in a single layer on the driveway. As the car and tractor run over them, they split the husk open and, after drying for several days, the nut is easy to remove. We spend a few minutes each day in October picking nuts off the driveway. We keep them in half-bushel baskets protected from rain so they can dry further.

**English Walnut** These fall free of the husk and usually drop over a wider period of time, at least in our experience, than other nuts. While most of the crop falls in October, we pick up a few scattered nuts from our young bearing tree in September and have found handfuls of newly fallen nuts on the ground as late as December. Keep them cool and dry like other nuts.

## Seeds

**Sunflower Seeds** Harvest the seed heads when the seeds are dry but before they begin to fall. Spread them in a dry, airy place to dry more completely before storing. Protect them from mice. You can rub the seeds from the heads and dry them on trays if you haven't room to hang the heads to dry. Keep the dry seeds in covered cans in a cool or cold dry place.

**Sesame Seeds** Store cold and dry.

**Popcorn** Leave the ears on the stalk until the kernels dry and harden. After picking, pull back the husks and hang the ears in a well-ventilated place to dry for one or two weeks. You can either store the ears whole, in cans or hanging in nylon stockings, or you can rub off the kernels and package them in tightly covered jars. Be sure the corn is good and dry before you seal it away, though, or it will mold.

If popcorn doesn't pop well, it has probably lost too much moisture. Add a few drops of water to the jar and wait a day or so. (It's the steam built up inside the kernel that makes it explode.) Keep stored popcorn in a cool dry place.

**Pumpkin Seeds** Store cold and dry. Dry them well before storing. Lady Godiva, Triple Treat, and Eat-All (a squash) have no hulls and must be kept cold in tightly closed containers. Seeds in hulls keep longer.

**Dried Soybeans** Air-dry for a week or two after threshing before you pack them away. Keep cool and dry and tightly closed.

**Peanuts** Southern gardeners have time to let the peanut vines grow yellow before they harvest the underground nuts, but in the north it's usually necessary to pull up plants at the time of the first serious frost, when they may still be green. Peanuts with soft, rather spongy shells and pinkish white skins are immature, and make poor keepers. The skins turn pink and later red as the peanuts mature and at the same time the shells harden. Leave your vines in the ground as long as possible to give the peanuts every chance to mature.

After pulling the vines, leave them in the sun for a few days to dry. Complete the drying by draping the vines over a sawhorse or hanging them from a rafter somewhere where air will get to them but rain will be kept off. Two to three weeks of this treatment should reduce their moisture content from an approximate 40 percent at harvest to a level of about 10 percent, just right for storage.

Keeping peanuts tightly closed in cans or plastic bags tends to encourage mold. It's better to seal them in sturdy paper or cloth bags. In the South, peanuts are commonly tied in burlap bags and hung from rafters or high hooks in a cool, dry, airy place.

## Wheat and Other Grains

Whole grains will remain in good condition for two or three years if kept cool and dry and protected from insects. Many people, though, prefer to raise or buy only enough for one year so that they will always have fresh grain. Grains should be kept in a tightly closed container that shuts out light. Large lard cans are excellent. Cardboard boxes and plastic bags are vulnerable to rodents and insects. Cans should be raised several inches above a concrete floor to prevent the metal from sweating and dampening the grain. Spilled grain shouldn't be allowed to remain in the storage area, because it attracts rodents and insects.

Harvest wheat when most of the stalks have turned yellow. For a small patch, you can pull or cut the stalks, bundle them together, and lean the bundles against each other to form shocks. Grain will cure in shocks in the field over a period of several weeks.

Keep home-raised grain in cloth sacks for a month after harvest, turning the bags periodically to ventilate the grain, before you put it in cans. Grains shouldn't be tightly enclosed in cans immediately after combining, for they still contain a good bit of moisture then. Moist grain is likely to mold if sealed, so grain should be well dried before it is put in airtight containers.

There are many small insects that make a career of eating stored grain. Sometimes they or their larvae are already mixed in with purchased grain, or they may lie in wait in crevices of cans and boxes you have on hand. Should you find that insect infestation is threatening your grain supply, you can use heat or cold to attack the bugs.

1. Heat infested grain in the oven at 135 degrees F for 30 minutes. Then leave the grain in the closed oven for an hour after turning off the heat.

2. Put the grain in the freezer where three or four days at 0 degrees F should render the insects inactive.

Grains don't belong in the damp root cellar, but since freezing doesn't hurt them, they may be kept on porches and in garages and unheated rooms.

In his book *Small-Scale Grain Raising,* Gene Logsdon suggests the following amounts of grain to keep the self-sufficient family well supplied for a year:

- Wheat: 1 bushel

- Corn: 2 pecks
- Popcorn: 2 pecks
- Soybeans: 4 pecks
- Buckwheat: 1 peck

## Sauerkraut

Sauerkraut made from summer cabbages must be frozen or canned to keep it from spoiling during the warmer months of early fall, unless you have a springhouse or an unusually cool root cellar to keep it in. Freezing preserves its flavor and texture better than canning, but most people need the room in their freezers for other foods. Canning works well, but it's just one more process to get done at a time of year when the days are already too short for all you want to do. The best arrangement is to grow your kraut cabbage as a late crop. If your cabbage matures in October and you make your kraut then, weather should be cold enough to keep the kraut in good shape by the time the two- to four-week period of fermentation is over. And unprocessed "raw" kraut tastes best of all.

You can cover the crock of kraut securely and keep it in a shed or garage until the weather gets cold enough to freeze it solid. When the kraut is cold but still unfrozen, bring it into the root cellar to keep it for the winter, or into an unheated room. It keeps well at between 30 and 40 degrees F. The cabbage shreds should remain submerged in the brine, so keep a clean plate weighted down with a rock or water-filled jar on top of the stored sauerkraut.

Kraut tastes best when fermented rather slowly at a temperature of 59 to 68 degrees F. At 86 degrees, cabbage ferments rapidly but the resulting sauerkraut will have a mushy texture and a sour flavor.

## Pickles

Brined crock pickles need no packaging or canning when made in late fall. For the best cucumber pickles, use fresh cucumbers. The small amount of sugar that cukes do have is lost during the first few days after picking. That natural sugar is needed to encourage the growth of the lactic bacteria responsible for the fermenting of the pickles. Green tomatoes are often more plentiful than cucumbers in the fall, and they make excellent brined pickles too. Keep cured pickles in a cold root cellar or pantry, between 32 and 40 degrees F if possible.

# Meat

Fresh meat is extremely perishable, but cured meats, especially hams and bacons, are often hung in cold cellars or attics for several months. Spoilage organisms become active at temperatures above 42 degrees F, so keep meats cold — 40 degrees and below. We've seen hams hung in root cellars, but a root cellar that is really damp enough for vegetable storage is likely to be too damp for cured meats. That moist air encourages mold. Hams keep better when winter temperatures are consistently cold. Periods of warmth like the proverbial January thaw encourage spoilage.

If you raise your own pork and have it butchered by someone you trust, or do the job yourself, you can omit harmful nitrates and still have good meat. A country butcher has prepared our hams for us for the past four years and each ham and bacon has turned out well without the use of nitrates.

# Cheese

A cold room or closet is a good place to keep cheese, too. A single shelf will do. Homemade cheese should be aged in a cool place to develop flavor. The ideal temperature is 45 degrees F. At lower temperatures the cheese will need several weeks longer to cure. At 50 degrees, on the other hand, different strains of bacteria will proliferate, and the cheese will have a different flavor. (In some European countries, especially in France, certain caves are prized for cheese curing because of their peculiar bacterial flora that impart a distinctive flavor to the cheese.)

Cheeses should be coated with paraffin or better yet with beeswax and turned every few days so that they ripen evenly. Homemade cheese is not at all uniform because it is impossible to exert complete control over all the variables: age of milk, feed the animal ate, temperature in processing, resident bacteria, and so on. Still, it is good — and interesting — to discover what flavor you get *this* time. Try curing cheese in different cold spots on your place to see whether the location influences the flavor. It might.

## Mushrooms

Now that you've got roots in the cellar, sprouting salads in the spare room, squash under the bed, grain on the porch, and nuts in the garage, can you spare another corner for some mushrooms? Home mushroom raising isn't exactly a surefire procedure, but when it works you have a supply of these capricious delicacies that is far fresher than any you could buy. And if you can provide the right conditions, you could even be plucking mushrooms in midwinter.

There are three ways to grow mushrooms at home. The first is free and ridiculously easy. Simply bring in an oyster mushroom log. You must know your mushrooms, though. If you're a reasonably experienced mushroom forager, you're familiar with the oyster mushroom — a white or cream-colored wavy-edged fungus with an off-center stalk and widely spaced gills. Oyster mushrooms grow on dead trees, not on the ground. (If you're unfamiliar with mushrooms, please don't use our brief description to identify your wild finds. Consult a good mushroom book and learn to identify the trees, too.)

Oyster mushrooms are found on dead elm and poplar trees (and some others) in the East, and on aspen, cottonwood, and maples in the West. It is often possible to persuade the dead wood, which is riddled with the mycelium (root growth) of the oyster mushroom, to bear more mushrooms if you keep it wet and fairly warm. First you must find a dead tree or downed log with oyster mushrooms growing on it. The mushrooms needn't be fresh — just whole enough for you to identify. If you're lucky (it doesn't happen *every* time) new ones will grow from the wood you bring in. Cut a manageable-sized section of log from the tree, or saw the tree into logs and bring in several logs. Slice off one cut end of the log evenly so that it will stand upright. Put the log in a pan of water and keep it in a warm place with a temperature in the high 50s or 60s. A corner of a basement or a large harvest kitchen would be just right. Keep the water in the pan replenished and start to look for fresh mushrooms in three to four weeks. When you've picked off all the mushrooms, keep the log moist for another month or two. Sometimes you can get a second harvest. And when the log's exhausted, there's the rest of the tree out there in the woods.

A second indoor mushroom growing method will allow you to raise mushrooms in your living room. Simulated logs made of compressed natural materials inoculated with special mushroom spawn may be purchased by mail from at least one seedsman. Growing directions are sent with the "preplanted" logs. Both oyster and Shiitake mushroom logs are

Oyster mushrooms grow on dead trees, and can be brought indoors, tree and all, for later havesting.

available. While not inexpensive, one log costs no more than a night out at the movies for four people, and it's a fun way to keep fresh food perking in odd corners of your homeplace.

Finally, if you have a damp cellar corner or shed that can be kept at around 52 to 65 degrees F, you might be interested in the more ambitious project of growing the common white market mushroom, *Agaricus bisporus,* in trays of compost. The procedure here is to make a special compost, pack it in deep wooden trays, and spread mushroom spawn (comparable to seed) over the surface. When the threadlike root growth has spread over the surface of the compost, you top it off with a layer of soil mixed with peat. Mushrooms begin to appear about three weeks after this.

The trays must be sprinkled often to keep moisture high, and the temperature should stay within the range mentioned above. The same conditions that grow good mushrooms are also conducive to the growth of competing molds and fungi, so sanitation is quite important. You'll need more detailed directions than these to grow mushrooms, but since extensive home mushroom growing is beyond the scope of this book, we refer you to Jo Mueller's book *Growing Your Own Mushrooms.* We simply want to point out how useful those odd corners under basement steps and in furnace and laundry rooms can be.

## Eggs

If you keep chickens, you've no doubt lamented as we have that you're too busy to do much creative baking in the summer when eggs are plentiful, and in winter when you have more time, the hens aren't laying! If you have a corner where the temperature can be persuaded to stay between 35 and 40 degrees F, you can keep some eggs to use during the hens' winter vacation.

The idea is to keep the eggs cool and somehow seal their pores or protect them from air. One time-honored way to stash away eggs is to pack them in crocks full of waterglass — a thick, slippery substance that effectively encases the egg and keeps out air. Some of our older relatives who grew up on waterglassed eggs tell us that the sensation of reaching into that cold thick gel to fish out an egg is not one of their favorite childhood memories, but the stuff does keep eggs in usable condition for up to five months.

You make up the waterglass solution yourself by mixing a pint of sodium silicate (available at your drugstore) with nine quarts of boiled, cooled water. Scald your crock or jar with hot water, pour in the goop and then carefully put the eggs in, always keeping a good two inches of waterglass above the top layer of eggs. Keep the crock in your root cellar or cold pantry and add boiled, cooled water as needed during the winter to keep the eggs well covered. (The waterglass level will go down as you use up the eggs, and as some water evaporates.) Never wash eggs before putting them away. Washing removes the natural protective coating that helps to prolong their storage life.

The staff of *The Mother Earth News* did a study on egg storage methods and concluded that unwashed fertile eggs will also keep for several months when stored in the following ways (in the study, eggs treated by the last three methods were kept at 65 to 70 degrees F):

1. Keep eggs cold (35 to 40 degrees F) and sealed in a covered container.
2. Coat the eggs with lard.
3. Pack the eggs in a crock of lard.
4. Submerge the eggs in a solution of 16 parts water, 2 parts lime, and 1 part salt.

Eggs packed in wet sand or sawdust spoiled more quickly than those that were untreated and left at room temperature.

## Potting Soil

Do you have room for one more thing in that cellar of yours? As you lay away good food for winter, remember that you'll need garden soil and compost in late winter to make potting soil for next year's spring seedlings. For most of us, some forethought is necessary to bring these seed starting necessities in under cover where they won't be rain-soaked and frozen when we need them. The root cellar or back porch is a good place to keep a bucket or two of compost and garden soil. And so the year turns, one season feeding into the next.

# SECTION FOUR

## Food Cellars for Everyone

# 12

# Trenches, Keeping-Closets, and Other Vegetable and Fruit Hideaways

The greatest fine art of the future will be the making of a comfortable living from a small piece of land.

Abraham Lincoln

Here are some simple ways to make all that garden goodness last for winter eating. Some of these tricks are hundreds of years old, but they're still used because they work. Others make use of contemporary scrap materials and a few use plastic sheets, but all are low-technology methods that require no fueled energy to maintain. The nice thing about these kinds of decentralized food storage is that they'll work for almost everyone. Even if you live in a rented house or grow your vegetables in a community garden or in a small backyard plot, you can build a clamp and dig a pit and stow some squash in the attic and eat well in a January blizzard from your own efforts. If you're still saving up the cash to put in an underground root cellar or an insulated room in your basement, you can store plenty of vegetables right now while you make your plans.

When we told a friend that we were working on a root cellar book, he said "Well, I live in a slab-based house on a flat piece of land. No basement, no hills. How can *I* store root vegetables?" Our answer is in this chapter. Having fresh food to eat for most of the year doesn't take a lot of land or equipment or dollars. Not even a whole lot of time, really — just some ingenuity and a ready acquaintance with the shovel.

You'll adapt these suggestions to your own situation, of course. You might want to try making comparisons between carrots stored in the row and in a pit, for example, to decide which method works best in your garden, for your family. Make notes on the results and experiment again next year. Start small and make a game of it. You might decide, as we have, to use both methods. If you're presently turning to frozen and canned vegetables in October or November, try one or two of these small-scale root cellaring tricks and see how much longer you can extend your fresh-vegetable season. Keep trying, keep experimenting, and before long you'll wonder why you ever put carrots or beets or onions or squash or sweet potatoes or cabbage on the shopping list in February, or even in April.

## Outdoors

### Garden Row Storage

You couldn't ask for a simpler way to keep vegetables. Certain root vegetables will stay in fine shape all winter long when left right where they're growing and covered with an 18-inch-thick layer of mulch, which may need to be held down by corn stalks or scrap boards. If you apply mulch before the ground freezes hard, it will often stay soft enough to dig for an additional month, and then all you must do to get fresh vegetables is to fork off the mulch, pry out a few roots, and replace the mulch on the row.

Soil in raised beds often freezes later than regular garden rows, and thus permits longer harvest of wintered-in crops. *Homesteader's News* founders Norm and Sherrie Lee of Addison, New York, report that they were able to pull carrots and beets from their well-mulched, intensively planted, raised beds well after conventional garden rows had frozen solid. The light, friable soil of a raised bed releases the vegetables more readily than heavy clay, and the fact that the beds are higher than the surrounding ground permits good drainage of both rain water and cold air.

Any vegetables you don't eat before the ground freezes closed will keep until spring, when they'll be a special treat. There's nothing to compare with the fresh carrot you munch on while working in the garden to get ready for a new season of growth.

Sometimes, though, mice work their way under the mulch and systematically destroy a whole planting, leaving a neat row of bare carrot-shaped holes – a spooky sight in early spring. You can discourage these little thieves by spreading a strip of hardware cloth or old screening down the row before putting on the mulch. And here's another trick you might want to try: Put a strip of sheet plastic down the row on top of an 8-inch layer of leaves or hay and cover the strip with another 10 or 12 inches of mulch. The plastic helps to keep the bottom leaves dry, so digging is easier without a frozen mat of soaked leaves to chip away.

## Garden Row Storage

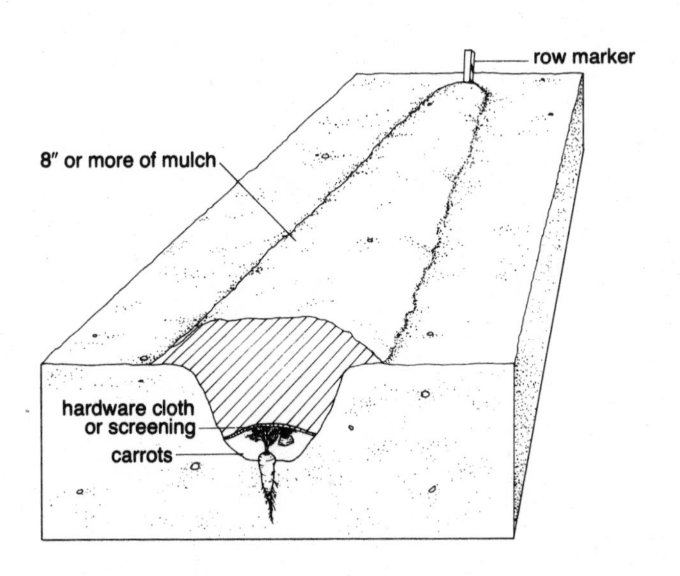

Garden row storage works well with kale (but don't cover it), carrots, parsnips, salsify, Jerusalem artichokes, and often turnips in the north. Add beets, cabbage, winter radishes, celeriac, and so on in the mid- and upper-south. Be sure to mark the rows with stakes so you'll know where to start digging. We know a northern gardener who dug up her carrot row and replanted the best roots in a deeply worked small square corner of the garden near the house, covering them in the usual way. They were easier to find, quicker to uncover, and more efficient to dig, she reported.

### Hay Bale Fortress

Build this to extend the life of otherwise perishable vegetables like lettuce, escarole, tomatoes, or to keep a prize broccoli or chard plant going even later into the winter than you normally could. It amounts to a temporary on-site cold frame. Just stack bales of hay about two deep around the plant or plants you want to protect, and cover the opening with an old storm window. On really cold nights, toss straw-filled bags or an old rug over the window. It's worth the trouble when you can pick your own fresh lettuce for Christmas dinner.

## Hay Bale Fortress

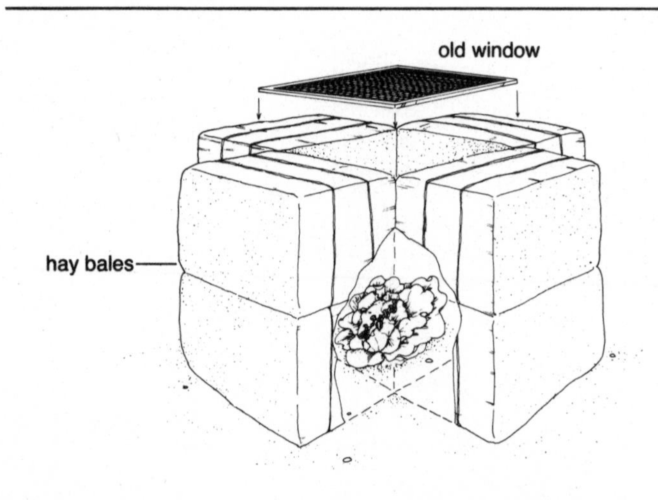

## Tents

Leafy vegetables protected by tents will often give you a good extra month of fresh picking — just when you need them most, during the cold days of early winter. Some of our favorite city gardeners have kept themselves in Swiss chard most of the year with a simple improvised tent. Use clear plastic for these tunnellike tents. Doubled cleaners' bags will last a few weeks, or use heavier plastic in which manufactured goods are often wrapped. You'll need a framework to support the plastic. Stiff wire bent into a U-shape and poked into the ground at two-foot intervals in the row works fine. Or you could use bent green twigs from tree brush, with both ends buried in the ground. Stretch the plastic over the framework and anchor it down at the sides with earth or rocks. A continuous band of soil is better because there will be no gaps for the wind to catch and tear the plastic.

### Plant Protection Tent

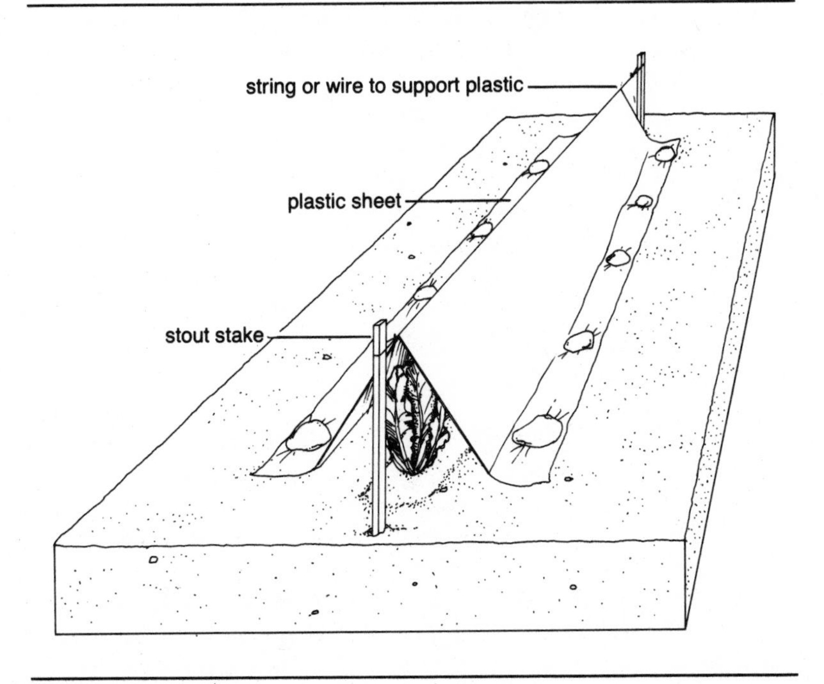

You'll need to ventilate your tunnel-tent, or the vegetables may cook when the sun shines. The open ends make good vents, but during very cold weather you might want to close over all but the bottom few inches of the tent ends at night. Don't leave the vent open at the top, or all the warm air will rise and drift away.

For more rigid, permanent in-the-row protection, you might find one of the portable glass or plastic covers useful (Guard 'n Gro and others). Or make a portable cold frame from an old storm sash and attach a carrying handle at each corner. You can set this plant protector right over the portion of the row from which you want to keep on harvesting.

Broccoli, chard, escarole, endive, spinach, and turnip greens are good candidates for tent treatment.

## Mounds or Clamps

Mounds or clamps have been relied on for centuries to feed land-based people through the winter. (A clamp is simply an above-ground mound of vegetables set out in the open and covered with earth and insulation.) These vegetable storage mounds have several advantages: They are not difficult to construct, they take very little material, and a whole row of mounds will fit handily in even a small backyard.

Mound storage isn't practical for the far north or for areas where winter temperatures average much above 30 degrees F.

Although making a winter-vegetable mound is not an arduous task, it is a job that must be done right. There are a good many mistakes you can make if you don't know just what is important in making a good mound. So take it from Mike, who grew up eating potatoes from clamps his mother built: here's what you need to do to build a good, workable clamp:

**Choosing the Site** Build the clamp on land that is dry and well drained, preferably slightly elevated with a natural slope for drainage, but certainly not a spot where ground water is high or where water tends to puddle. Don't worry about not having a slope — you can remedy that — but if you have one, use it.

Soil quality, while not a factor you can change quickly, does influence the effectiveness of the clamp. The ideal soil to use is a light, sandy loam, because it sifts well between the vegetables and is easy to dig into.

Heavy clay which breaks up only into rough clods doesn't cover vegetables well and packs hard, making digging difficult. The soil on which you build the clamp should be free of raw organic matter which could spoil your produce as it decomposes. And if you built a clamp last year, put this year's clamp in a different spot if at all possible, to avoid building up healthy populations of the kinds of pests that thrive on storage vegetables, especially potatoes. If rotation isn't practical for you, then just clean off and compost all the old straw, spoiled vegetables, and such from the previous year's mound and air out the spot for two days before rebuilding on it.

Another site factor to consider is protection from prevailing winds. Wind that whips unimpeded across three cornfields before it hits your place will put a heavy chill on anything that sticks up out of the ground. If you can, construct your clamps in a spot that is sheltered from wind by trees, buildings, hedges, or fences; or, if necessary, devise a hay-bale or burlap-screen windbreak. The mound shouldn't be too close to sheds or other buildings, though, or rodents may find it an easy mark.

**Laying the Foundation** Rake leaves and garden debris from the spot you've chosen for your clamp and dig out 8 to 12 inches of soil, saving it separately to cover the pile later. Pack a mat of straw, hay, or leaves about 3 inches deep into this shallow pit. (Some gardeners start the mound right on the ground surface without digging out any soil.)

**Stacking the Vegetables** Carefully stack your vegetables on the hay or straw mat, aiming for a cone-shaped pile. You can make the pile two or three feet high if you like, but it's important to remember that once opened, the clamp must usually be emptied because it can be very difficult to close over securely in freezing weather. The vegetables you remove would keep well in a cold, damp place, but if you have only a warm house to bring them to, they'll spoil soon. So you'd probably be wise to make several small clamps rather than one big one.

Can you mix the vegetables? Yes, indeed, in fact this is a practical idea that will help you to get more value from your free cold-storage arrangement. When you open up a cache of assorted good keepers — carrots, turnips, rutabagas, beets, and potatoes — you have ready the fixings for a variety of meals. But don't store fruits and vegetables together. When mixing vegetables, you might find it helpful to remember that the highest temperature in a clamp will be toward the top.

As you arrange the vegetables, pile them around a small bundle of

## Clamp or Mound

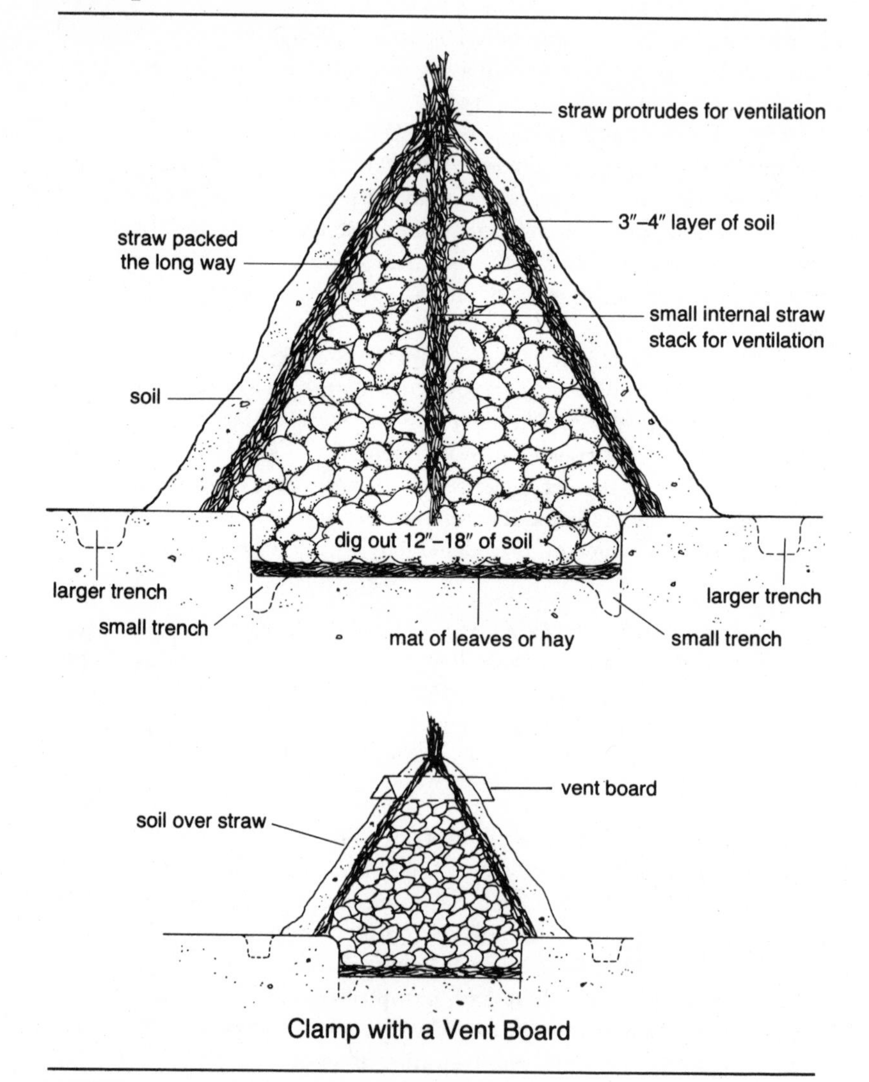


Clamp with a Vent Board

straw or brushwood, or a perforated pipe poked into the center of the heap. This acts as a ventilating shaft, admitting fresh air from above to circulate among the stored vegetables. Piled-up vegetables often tend to heat early in the storage period, so ventilation is necessary to draw off this heat and prevent spoilage.

Before you begin to cover the clamp, dig a shallow trench all around the pile of vegetables (or do this before stacking them). This provides extra drainage insurance.

**Covering the Clamp** Now you're ready to complete the clamp. Pack a generous blanket of straw or hay over the whole surface of the mound of vegetables. Leaves may be used but they're difficult to anchor in place unless you stuff them in bags — and then you must be sure there are no exposed gaps. Stiff straw is the best material to use, because it leaves plenty of tiny air spaces and doesn't mat down. Position the bundles of straw the long way with their ends at the top of the pile. Straw that comes in bales is often short. The straw that Mike's mother hand-sickled from the rye field was perfect for this purpose — long, strong, airy, and springy. Use what you have, but make it a good thick layer — 6 to 12 inches. Form the straw into a thin point at the top of the pile — a continuation of the natural ventilating system. You can cover the straw with a board to keep mice out.

Then shovel on a three- to four-inch outer layer of dirt, covering the whole pile except for the tip of the straw vent on top. Build the dirt layer up gradually and pack it down well with the back of your shovel so it's firm. Finally, make a second trench around the mound to drain off rain and snow melt. Don't surround the pile with mulch, or mice may decide you meant it just for them.

European gardeners sometimes use boards to form a vent instead of straw or fine brushwood. In this case, they nail two boards together forming an inverted "V" and place this frame over the vegetable cone before covering the vegetables with straw and soil. Both ends of the board should be long enough to extend beyond the edge of the straw-dirt cover on both sides of the mound.

You don't want to make your clamp covering completely airtight. Never cover the pile with plastic. If ice forms on the mound, the lack of air exchange can make the vegetables heat up. An ice-glazed mound is also difficult — and sometimes impossible — to break into.

Unlike refrigerators and freezers, clamps don't come with any guarantees. You can't control the temperature and the thing may be a bit unwieldy to dig into when the weather's fierce. Sometimes you even need a pick to pry it open. But you never owe anyone a penny on a clamp, and it never raises your gas or electric bill, either.

Vegetables to store in clamps are potatoes, carrots, beets, turnips, winter radishes, cabbage, rutabagas, celeriac, and parsnips. Fruits are apples and quinces.

## Buried Refrigerator

A dead refrigerator can be turned into a fine vegetable storage vault if you're willing to dig a hole for it. We tried keeping carrots in an above-ground frig insulated with hay packed around it and on top of it, but when the thermometer went below zero, the carrots froze. We should have put more hay on top of the box than the eight inches or so that we used — whole bales would have been better. Burying the frig is the thing to do. That way you get the advantage of the more constant and moderate temperature of the earth under the soil surface.

**Buried Refrigerator**

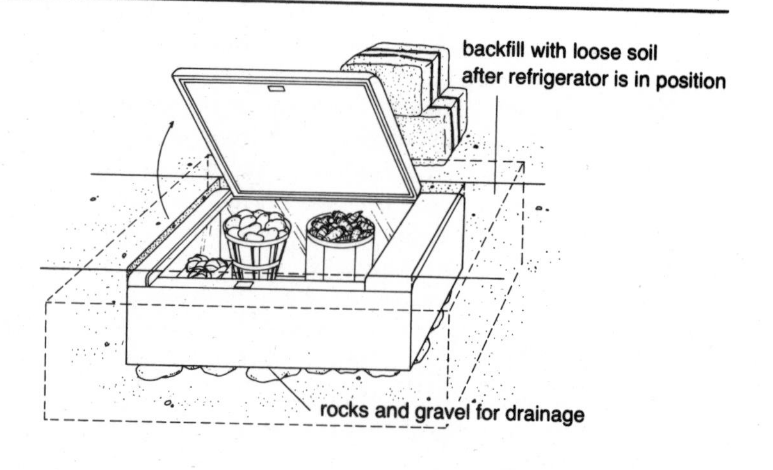


If you have an unused refrigerator you'd like to make into a garden root cellar, remove the motor and shelves and crisper drawers. Be *very* sure to knock off the lock so that no child can be accidentally locked in the box. Dig a big hole a foot each way larger than the refrigerator and toss some rocks in for drainage. Ease the empty frig down into the hole and position it so it's on its back. The door will now swing up like a chest freezer. The top surface of the buried refrigerator should be at or slightly below soil level. If you want to store several kinds of vegetables in the frig, you can leave one or two shelves in for dividers.

Fill in between the refrigerator and the ground with loose soil. When you've packed your vegetables away, heap bales of hay or bags of leaves on the refrigerator lid to keep winter cold from descending into the buried box. It's a good idea too to cover the lid with scrap boards or metal roofing (we used a heavy piece of ribbed rubber matting) to prevent water from seeping into the joint between the lid and the body and freezing them together.

You can keep apples, potatoes, beets, carrots, turnips, celeriac, kohlrabi, and rutabagas in a buried refrigerator. You might find that you need to run a small vent pipe into the refrigerator to admit fresh air.

## Trenches

Trench storage works well for leafy vegetables like celery, Chinese cabbage, cabbage, and chard, which might spoil if heaped in a pile. You'll need a shovel, dirt, and straw as you did for the clamp, but you'll get to do a little more digging on this one. Remember, you're counting on the moderating temperature of the deep-down soil as well as the insulating effect of the batts of hay and leaves you pile on top.

Dig your trench before you uproot the vegetables that will go in it. Save the fine topsoil to sift around the roots. Make the trench about 2 feet deep and as wide as necessary to accommodate the vegetables without crowding them. Then spade up your leafy vegetables, root and all, and replant them in the trench. Cover the roots with fine soil and water them, using a long-spouted watering can so you don't douse the leaves. Wet leaves will rot quickly in a trench. Place boards over the trench to serve as a roof for the vegetables and top the boards with a 12- to 18-inch layer of hay, leaf-stuffed bags, or corn stalks. Some gardeners like to nail two boards together at right angles to form a little peaked roof that will shed water better than the flat roof.

While temperatures in a clamp are highest at the top of the vegetable pile, trench temperatures are usually highest at the bottom of the trench closest to the earth's constant moderating influence.

Cabbage is often treated in another way — positioned root-up, with the head resting on a three-inch layer of straw, leaves, or other padding in the bottom of the trench, then covered with boards and more hay or other insulating material over the boards. Trench storage is especially good for cabbage since it sometimes affects the flavor of other foods if stored close to them.

## Storage Trench

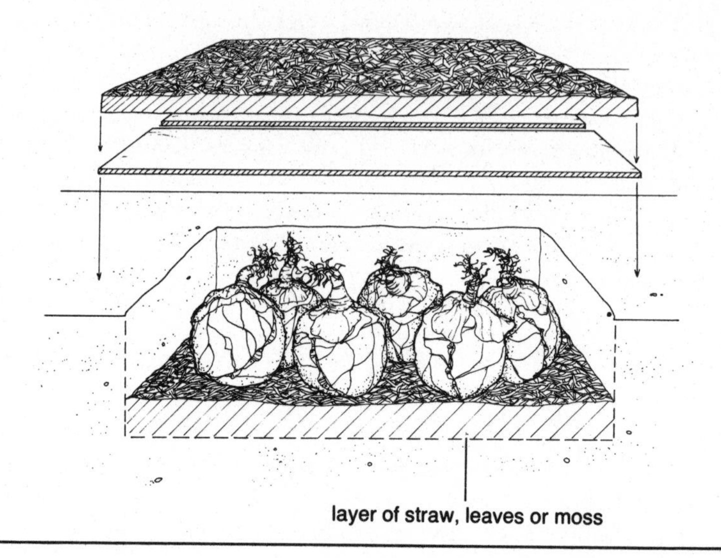


Another trenching arrangement involves placing a triangular frame of scrap wood at each end of the trench and nailing pieces of lath to these frames to form a sort of triangular crate. Pile a layer of straw over the frame and then pack on a six-inch covering of soil over the sides of the frame. Leave the ends open for ventilation, and stuff them firmly with dry straw.

### Buried Barrel, Drain Tile, or Covered Metal Can

The underground barrel root cellar is an old favorite, but large drain tiles are probably cheaper and easier to find than barrels these days. If you have an old barrel, though, here's how to install it as a vegetable keeper.

Get out your shovel again. If the ground is rocky, you may even need a pick to pry out the stones. This root cellar business does wonders for the appetite! In a shady spot, dig a hole deep enough to accept the barrel positioned at a 45-degree angle. Scrape out the loose dirt and toss a layer of rocks into the bottom of the hole. Lower the barrel into the hole, then firm the soil around it and pack a good two feet of dirt over and around the barrel so you end up with kind of a hunched mound.

Now you can fill the barrel with vegetables. If you're putting in an assortment of vegetables, include some of each kind in every layer you put in, so you don't have to eat your way through all the turnips to get to the beets. As you put the vegetables in the barrel, tuck straw, hay, leaves, or moss around them to insulate and cushion them. Stop loading before you get to the top of the barrel. Fill the last four or five inches with an armful of hay or straw for insulation. Then clap on the wooden lid, jam a foot-thick wedge of straw over the lid, and press a board over the straw. Finally, roll up one of those big stones you had to pry out while you were digging and wedge it against the board to keep the whole arrangement together. One of our respected country consultants tells us that his family always waxed the oak barrel in which they stored their underground vegetables to make it more waterproof.

To bury a big clay drain tile flue liner or a galvanized lidded can, just dig a hole straight down, not at an angle, large enough to accept the container. Pour an inch or two of gravel in the hole and then put the tile

## Storage Barrel

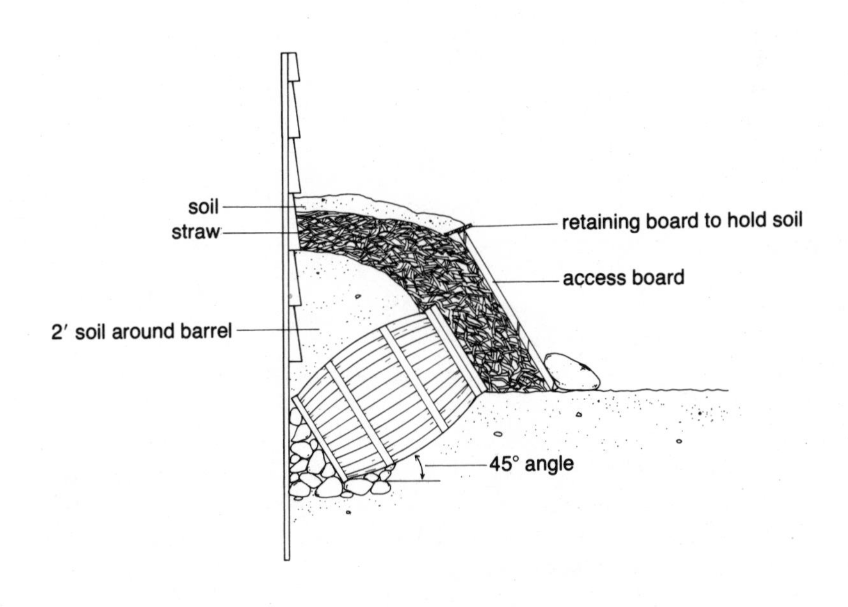

or garbage can in place. (The tile would be our first choice. Clay insulates better than metal and won't rust out as the metal can eventually will.) If you're at all unsure of the drainage, put a layer of rocks in the bottom of the tile before filling it with vegetables. You can set baskets of produce right in the tile and lift them out. The tiles may be obtained in 18- by 20-inch or 24- by 24-inch sizes.

## Drain Tile Storage

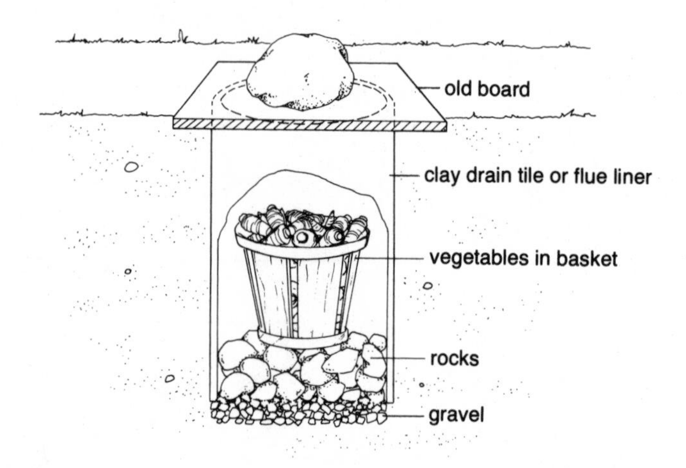


## Root Cellaring in a Box

Doesn't all this digging make you feel virtuous and fit? Here's another way to store your vegetables without using any of the expensive, oil-based kind of energy. Dig a hole not too far from the kitchen door, big enough to accomodate a box about two by four feet. You can make the box larger, but we wouldn't make it any smaller than two by three feet. Construct a simple wooden box out of scrap lumber to fit in the hole. Line the box with hardware cloth to keep out gnawing rodents. Arrange the vegetables on a bed of hay with hay, straw, leaves, or moss

## Root Box

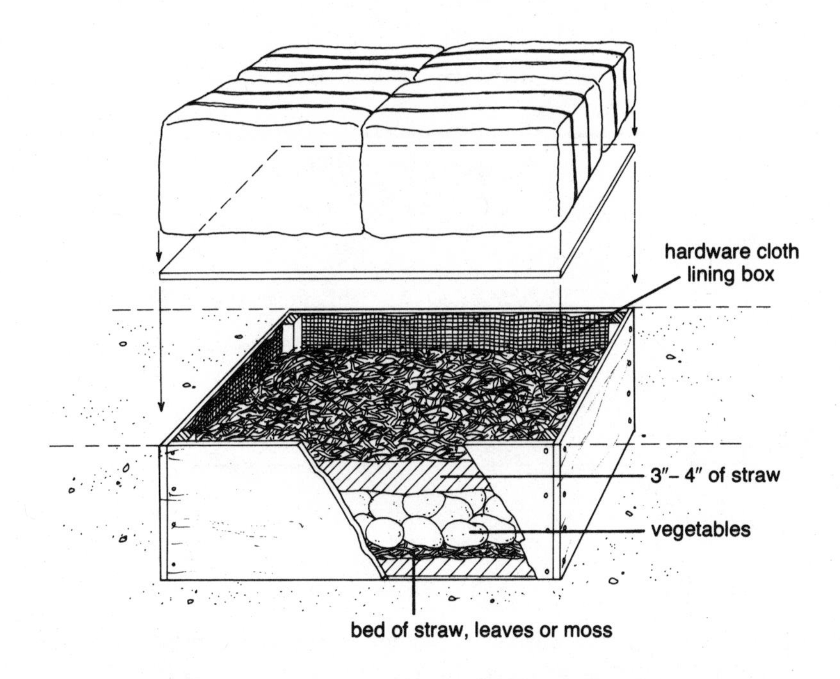


packed between the layers. Cover with a three- to four-inch layer of hay. Top with a hardware-cloth-lined lid — not hinged, just laid on top. Put hay bales on top of the lid and enjoy visiting your underground store when the roads are frozen and more snow's on the way.

### Earth Pits

The earth pit is even easier to arrange than the wood-box safe, but if rodents intrude you'll need to line the pit with hardware cloth. Here's how to make one.

Choose a shady, well-drained spot within shouting distance of the kitchen door. Dig a pit about four feet wide and two feet deep. Pack the earth around the pit to form a sloping collar that will direct water runoff away from the opening, then dig a trench all around the pit to channel off rain water. Shovel a two-inch layer of dry sand or sawdust into the

## Earth Pit

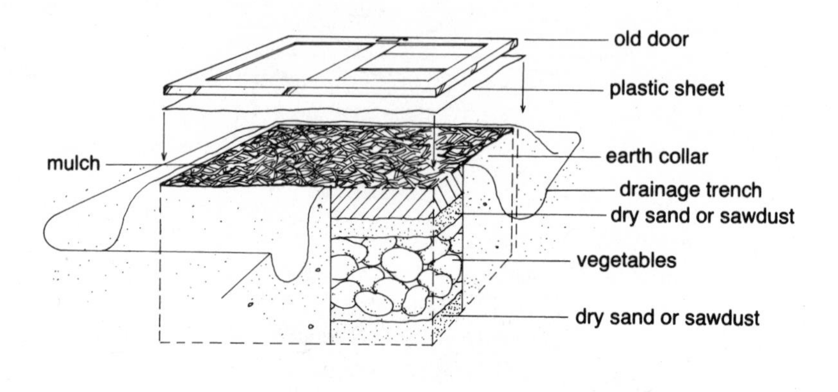


open pit, and lay the vegetables carefully on the sand, sifting more sand or sawdust between the vegetables until you've built up a layer of vegetables a foot high. Top this with another two or three inches of sand or sawdust, then a foot-thick blanket of hay, leaves, straw, and so on. Spread a layer of plastic sheeting over the straw and place an old door or several planks on top to hold it down. You'll feel like you're earning those vegetables as you pry your way in to find them, but they should be in good shape when you get there.

These pits are good for potatoes, carrots, turnips, beets, parsnips, and rutabagas. You can keep apples in earth pits too, but not with other fruits and vegetables.

### Oat Bins

Who'd think of hiding a melon in the oat bin? This is where Grandfather often put the last cantalopes of the season, to ripen while he did the fall chores. A bin of barley, wheat, or rye will work just as well. The grain admits some air but cushions and insulates the fruit — a neat arrangement. Old-timers often kept cured hams deep down in grain bins, too.

## Outbuildings

You may not have a basement or a hill to dig a root cellar into, but if you have a garage or other outbuilding, you can make it work for you at harvest time. Your climate will determine how long you can store your produce in an unheated outbuilding. We've used our shaded garage and shed to store apples, potatoes, onions, nuts, beets, and some other fall vegetables during the first few postharvest weeks when nighttime temperatures dropped below freezing but not too far below. We've also stored vegetables and apples well into winter in small heaps covered with bags of leaves in an unheated garage in central Pennsylvania. When we spent a year in Wisconsin, in our innocence we left a late picking of apples in an insulated camping cooler in a small yard shed. The apples froze, but even though they were mushy and icy in January, they were very sweet and so good that we ate them all.

You can carry vegetable protection one step further by packing your produce in cartons, coolers, boxes, bins, and so on with an insulating layer of sawdust or hay on the bottom, sides, and between layers. Thus jacketed and stored in a garage, they'll keep through December at least in our area — and often into January. But soon after New Year's Day we find ourselves quoting that old saying: "When the days begin to lengthen, then the cold begins to strengthen." It seems to be true, so any vegetable kept outside under wraps should be brought to a more protected spot by early January — an unheated porch, perhaps, or a cold attic. In far northern states, you'd need to keep a watchful eye on insulated outside edibles in December and bring them in when temperatures head for the low teens.

If your garage or shed has a dirt floor, you could excavate a pit right there for vegetable storage. With walls and roof protecting the pit from ice and snow, you'd have less trouble breaking into it. One man dug a shallow trench along the wall of his shed to store cabbages. Another clever gardener mentioned in *Keeping the Harvest* by Nancy Thurber and Gretchen Mead, buried a metal wall locker in his garage for stashing vegetables away. This is a particularly neat solution because the metal sides keep out mice and rats, a locker is usually already ventilated, and lockers often are available for a song from salvage sales. One enterprising family with a garage built into a hill used a necessary wall repair job as an opportunity to burrow another few feet into the hill, through the inner garage wall, to make a winter vegetable closet.

We know an apple grower who hangs baskets of his fruit from the rafters of the head-high basement space under the front porch of his log cabin. He twists wire coat hangers into hooks for this purpose. The hung-high fruit is safe from flooding, pretty safe from mice, and easy to get to. You could do the same thing with beets or turnips if you had a similar space.

Another family we stopped to visit had an unused underground concrete cistern they intended to convert to a root cellar. They'll need to run pipes into the former water vault to ventilate it, keep a ladder in it for access, and add a more convenient trapdoor on top, but we'd guess that with a little tinkering their ready-made root cellar should work out very well. Being able to use a facility they already have helps to make up for the inconvenience of climbing down the ladder to reach the stored food.

These are just a few of the many imaginative arrangements that could be worked out. Look around *your* place and list the nooks and crannies where you might tuck a carton or a cooler of carrots. Then figure out how you could fend off the frigid air from that place a bit more effectively — by hanging old blankets on an inside shed wall, mounding up leaf-stuffed bags, or exhausting some house or dryer heat into the garage through a vent or window. Although your carrots, beets, and apples might not last until spring in a shed, they could last well into the winter, depending on how well you can cushion them, and while you're enjoying those extra weeks of vegetable independence, you can be figuring out how to extend them.

## House Nooks and Crannies

The average house has at least one unsuspected area that would make a good root cellaring space. If you're lucky, you'll find several! Remember that you're looking for more than one set of conditions — a cold, damp space for potatoes and other root vegetables, and a warmish dry spot for pumpkins and squash, and perhaps a cool dry corner for garlic and onions.

Do you have an attic? That's a great place to winter pumpkins and squash. Most attics are unheated and it's usually easy to open a door or vent from the house to admit some additional heat from time to time if that is necessary. The only problem you might have with attic storage, providing the attic doesn't freeze, is that produce you put up there is easily forgotten. Write notes on your appointment calendar to remind yourself to use and periodically inspect your upper-level larder. It's always nice to find something you didn't know you had, unless that something is a mushy two-year-old squash.

The unheated room can be a real asset in winter. Perhaps it's a spare bedroom, an unused extra bath, a pantry, or just a big coat-closet. Keep the door closed for a day, put a thermometer in the room to give you an idea of how cold it gets, and then decide what kind of vegetables would keep well there (see chapters 8 and 9).

You can sneak several bushels of produce into a room without making it look like a warehouse, in case you need to use the room occasionally. Cover stacked cartons with an afghan, quilt, or attractive print sheet. Hide small quantities of vegetables under the floor-length skirt of a corner table. Line up interesting baskets full of sweet potatoes along one wall of the room. Toss a small throw rug over a filled crate.

The old house on our farm hasn't a single closet, but it does have a cold back room, exposed to the west wind, where squash, onions, garlic, and sweet potatoes keep quite well from November until March and often April. (During the few bitter cold days we usually experience every winter, we simply move the produce to an upstairs room where it's a few degrees warmer.) Since we heat the house with an airtight wood box stove, the kitchen and the upstairs bedrooms are comfortably warm. Other rooms, into which the heat doesn't circulate without the help of a fan, are cool and sometimes cold in winter. (The house is not insulated.) In homes like ours, where reliance on wood heat has replaced dependence on central heating systems run on purchased fuel, it's usually easy to find a gradation of cool and cold rooms that are useful for food storage. In old houses, where rodents can easily enter through the foundation, you'll need to keep an eye out for mice and set traps if necessary, if you have a lot of foodstuffs stored out in the open.

An enclosed porch is as good as an unheated room, and so is a cellar, of course. And don't overlook those odd spaces under the porch steps or in a breezeway or outside cellar entry. It's all very well to have a perfectly finished, efficient house, but some of these irregular little cubbyholes that might make an architect shudder will prove mighty useful for natural cold storage.

The crawl space under a porch or elevated house is dark, cold, and damp for most of the fall — good root storage conditions. When outside weather turns very cold, produce kept in crawl spaces will freeze without additional protection. Insulating the space and packing your vegetables in sawdust or leaves will help to extend the useful period of your crawl space larder.

A Virginia gardener wrote to tell us about his under-porch root cellar. He has a high screened porch on the northeast side of his house. There are steps down to the cellar under the porch, but the furnace makes the cellar too hot for food storage and the space is too small to partition.

So our gardener friend made a framed-in root cellar in the dirt area under the porch, right next to the basement steps. First he dug a hole in the ground to a depth of 18 inches. Then, using two-by-fours, he framed in a rectangular space 46 inches long, 42 inches high, and 25

## Under-Porch Root Cellar

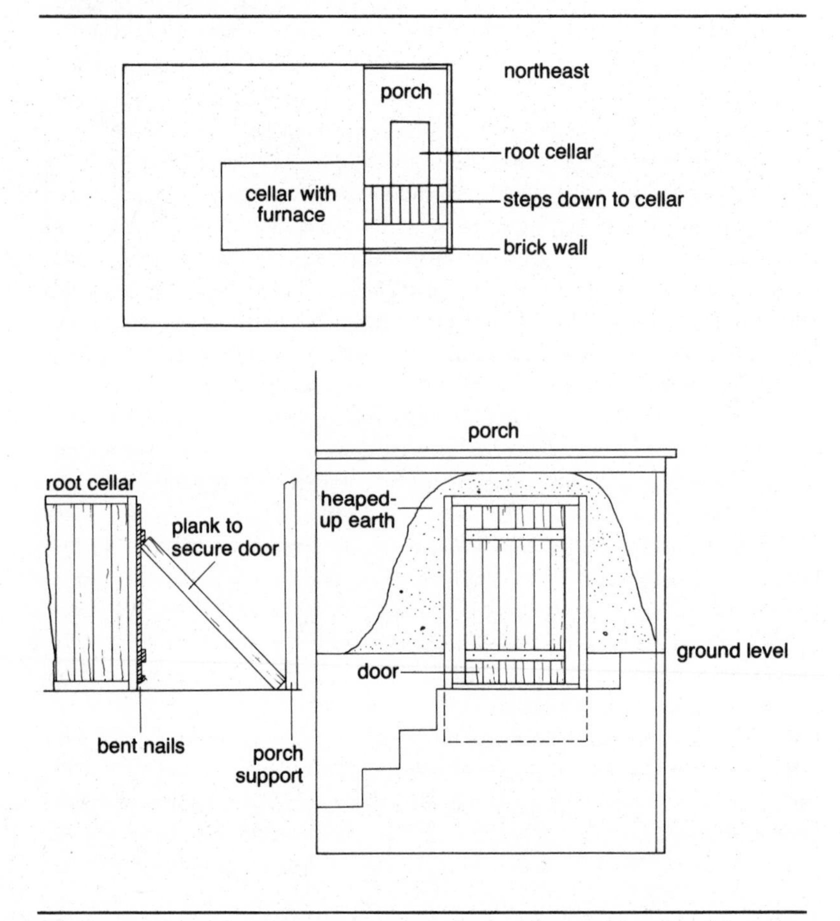

inches wide. He used old planks to cover the sides, ceiling, and floor of the little root cellar. Next he lined the walls with sheet tin to keep out varmints. Finally, he put on a simple door — just a plank frame, lined with tin, which he pins in place at the bottom with nails bent upward. A plank propped against an upright under the porch adds the strength of a buttress to keep the door closed. Soil heaped around the half-buried box helps to insulate it.

A closet on the north side of the house could be given over to fruit or vegetable storage. Wall space along basement and attic stairways might be fitted with shelves to hold small cartons of onions. If you have steps leading directly from your basement to the outside, the cold outside air descending through the passageway will refrigerate baskets of produce you set on the steps. You can even vary the conditions to suit different foods. It's usually coldest on the top steps closest to the outdoors, and warmer on the lower steps near the cellar. If the cellar door is exposed to the sun, or your area gets very cold, it might be worth your while to insulate the door, either with Styrofoam or fiberglass fastened to the inside, or hay bales or bags of leaves piled on the outside if it's a slanting bulkhead door. You can spread damp burlap sacks over the produce to increase moisture.

Basement window wells offer a small area but they're easy to fix up for vegetable storage. In an old home, thick basement walls may permit you to box-in an interior space, which you can cool by opening the basement window, and warm by admitting heated basement air through a vent in the box. Or, if you live in a newer house that has an exterior window well, you can place a box with a hinged lid in the window well, insulate around the outside of the box with shavings, straw, or whatever, prop open the basement window, and wall in the sides of the enclosure with scrap boards. In effect, you've just extended the basement area into the window well. Such an arrangement will allow you to reach through the open basement window to open the box lid and remove the vegetables you need for dinner. Mice like these cozy set-ups too, so watch for evidence of them or cover the box with hardware cloth if mice are already a problem.

Now that you've surveyed your homeplace, we hope you've been able to find several good storage spots that will keep you in fresh vegetables longer than ever before. And we hope you've had some fun improvising and experimenting, too, using materials at hand to come up with a good solution for your situation. What can *you* come up with? We'd like to hear from you. Would you let us know about any different storage arrangements you work out?

# 13

# Planning Your Root Cellar

Happiness belongs to the self-sufficient.

Aristotle

The kind of root cellar you build or adapt from what you've already got, will depend on the floor plan of your house, the lay of your land, and the prevailing winter temperatures and snow cover in your area. If you have an old house with an unheated, dirt-floored basement room you're all set. With slight modification, or in some cases none at all, you may have a near-ideal vegetable storage cellar. In a house with a heated basement, it may be possible to partition off an unheated corner for vegetable storage. If you prefer an outdoor root cellar, you can either dig into a hill or go straight underground and top the cellar entrance with a bulkhead door, patio, or porch. If you live in an area where winter temperatures are often below zero degrees F and snow cover is heavy, you may prefer climbing a ladder down a hatch into an under-porch storage pit to shovelling through drifts to get to a hill-cave root cellar.

If your winters are mild, with average temperatures much over 30 degrees F, you will not be able to achieve the most desirable low temperature in your cellar for the keeping of root vegetables. To compensate, though, these vegetables will probably keep very well in the garden row for you and the warm keepers like squash, onions, and sweets should do very well in a cool corner of your house.

Whether you're modifying an existing space or building from scratch, you'll want to consider the ground rules for a good root cellar when drawing up your plans. While it is not always possible to arrange a textbook-perfect root cellar on the homeplace with just the right moisture and temperature for each vegetable, it makes sense to understand the basic principles of root cellar construction. Then you can at least aim in the right direction. If you can't provide ideal conditions, don't despair. Plenty of people live quite well all winter from root cellars that don't precisely meet all of these basic requirements. On the other hand, if you are building a storage area from the ground up, you might as well do it right.

We're thinking, for example, of a house built for the pastor of a nearby church by his parishioners. The well-meaning builders put in a root cellar under the house, but they did not provide any means of ventilation, and the pastor, who was also an enthusiastic gardener, found that his vegetables didn't keep as long as they should. It's still possible to ventilate that cellar, but it would have been much easier to provide ventilation during initial construction.

You'll find plans and drawings for different kinds of root cellars in chapters 15 and 16 of this section. Before you start nailing or digging, though, here are some important factors to consider:

There are three basic conditions your root cellar should provide. The closer you can come to approximating these ideal conditions in your vegetable storage area, the better your vegetables will keep.

## Temperature

Temperature should be your first consideration. The largest group of storage vegetables and fruits (see chart, page 51) keep best in a cold place. The warmer temperatures required by squash, sweet potatoes, pumpkins, onions, and so on are easier to provide in the average home without making any special arrangements. So, although not *all* good keepers need cold, we'll concentrate here on low-temperature storage because that takes the most doing to arrange.

A thermometer is a necessary tool for a well-run root cellar. One with minimum-maximum readings will give you a good picture of the kind of temperature variations you're working with.

A good root cellar can both *borrow* cold and *keep* cold. You borrow cold by digging into the ground, where the temperature well below the frost level remains a fairly constant 52 degrees F or so. This works two ways. The deep-down earth temperature is slow to be affected by ultra-cold surface temperatures too, so your underground vegetables have an extra margin of protection against freezing. You also borrow cold by providing some means of admitting cold night air to the cellar, especially during the fall when days may be too warm to promote good keeping. Three effective cold air ducts are:

1. window,
2. louvered ventilator, and
3. exhaust pipe with plug or other means to close off the pipe when desired.

If you can maintain temperatures between 32 and 40 degrees F, you have an excellent storage place. A temperature range of 40 to 50 degrees will still permit shorter-term storage of root vegetables and apples and will keep onions and some of the short-lived storage vegetables like peppers, tomatoes, and eggplant in good shape for a month or so. In an indoor root cellar, the area close to the ceiling will be a few degrees warmer than space near the floor, so you'll have at least a small difference in conditions which you can use to advantage in placing vegetables with slightly different storage requirements.

Once you've cooled the food cellar, you want to keep it cold. Here again, you need some way of letting cold air in so that you can adjust the temperature if it begins to rise. Even more basic, the placement of your cellar will help you to keep it cold. The best spot for a basement root cellar is on the northeast side where the sun's warming influence is the weakest. Next best would be the northwest side, especially if trees or bushes shade the area from the western sun.

Most outdoor root cellars are dug into a north-facing hill or are underground on the north (coolest) side of the house. (You'll find exceptions to this rule in some of the northern states and Canada, where temperatures plummet well below zero in winter, and stay there. Some of the old-time outdoor root cellars in these areas are built on southern exposures so that winter access will be somewhat easier. Once winter sets in, it's plenty cold there even though these cellars get more sunlight than it would seem they should.)

In a basement vegetable closet, it's important too, of course, to keep out heat. Build the closet in a corner away from the furnace, heating ducts, and hot water pipes. If you can't escape the warming pipes, insulate them with a wrapping of fiberglass or one of the pop-on foam tubes.

Many underground root cellars are built with double doors that form an airlock or anteroom which helps to keep warm summer air out of the cellar and also to prevent undue chilling of the produce in severe winter weather.

Insulation also helps to maintain a stable temperature in the storage space. It won't help much, in a room you've simply partitioned off from a heated basement, to open a window to cool the room. Heat from the basement will warm up the space, and the warmth from the heated living area above will also influence temperatures in the cool room below. The solution is to insulate not only the warm inner walls, but also the ceiling and door, of a partitioned vegetable closet. Insulation should not, of course, be applied to the cold outside walls. Old-timers used sawdust, wood shavings left from planing, cork dust, and even cinders, straw, or dry leaves. More recently developed materials with higher resistance values include fiberglass, urethane, Styrofoam, mineral wool, and vermiculite or perlite.

Although brick and concrete walls look substantial, their insulating value is quite low. If your cellar temperature runs around 60 degrees F, you'll need the following thicknesses of material to keep the root cellar cold:

- Planer shavings: 4–5 inches
- Mineral wool: 3 inches
- Powdered cork: 3 inches
- Fiberglass: 3 inches
- Styrofoam: 2 inches
- Urethane foam: 2 inches

Urethane and Styrofoam are flammable. Cellulose fiber, which is used as loose fill, should have been treated with fireproofing materials; check the bag label. Sawdust and mineral wool used as wall insulation will settle over a period of time. Planer shavings are less likely to settle if they are tamped into place rather firmly. Shavings should be compressed to a density of about seven pounds to the cubic foot. Some kinds of insulation, fiberglass batts for example, have a vapor barrier attached to one surface. Be sure to apply the insulation with the vapor barrier on the outer, or warmer, side so that moisture doesn't condense on the insulation and spoil its effect. When using loose, dry insulation you can apply a sheet of polyethylene between the insulation and the warmer wall surface to keep out moisture that would otherwise condense at the interface of warm and cool surfaces. When insulation becomes wet it loses its effectiveness because it then has less trapped air and it often packs down and thus loses volume.

## Humidity

High humidity is the second requirement for effective root cellar storage. As you'll note in the chart on page 51, most root crops and leafy vegetables keep best at a humidity of 90 to 95 percent. Providing plenty of moisture helps to prevent these foods from shrivelling.

How can you measure the humidity in your root cellar? It's not as difficult as you might suppose. Simply purchase a hygrometer at a hardware store. The hygrometer measures humidity. Most of the successful grower-keepers we've met use a thermometer but do not actually measure humidity. The modest expense would be well worth your while, though, if fine-tuning your storage conditions would increase the amount of food you can put by for winter.

There are three ways to achieve the necessary humidity in a root cellar. For one thing, a cellar with a dirt floor will contain more natural moisture than one with a concrete or stone surface. In many older homes, one basement room was purposely left unfinished for just this purpose, even when "improvements" like furnaces were added to other parts of the basement.

We had such a room in a turn-of-the-century house we once owned. The furnace room had a concrete floor but one-third of the large basement area retained the original dirt floor. Potatoes, carrots, and apples kept very well for us there. We simply mounded up the potatoes on the floor and covered them with loose fine dirt. The room was, in fact, so damp that squash kept there spoiled early. In our new root cellar, we've left bare earth in a corner of the small basement area and have spread gravel on the packed earth. This helps to keep feet dry if the ground really gets damp. In addition, should the storage area need more moisture, we can easily sprinkle the gravel with water and its large surface area will evaporate the moisture readily into the air.

And that's the second way to keep humidity high — add some water, either by sprinkling the floor, spreading damp (but not dripping wet) burlap bags over the produce, or setting pans of water on the floor. Such measures are often necessary in the fall when you're first stowing the produce away. Basement root cellars are much more likely than dug-in root cellars to need added humidity.

In a very moist area you can often simply keep the root vegetables uncovered in bins and they'll stay smooth and firm. If your humidity reading falls short of the ideal, though, you'll often have better results if you use a third technique — packing the root vegetables, especially carrots, beets, and parsnips, in damp sawdust, sand, or moss to cut down surface evaporation. You can also encase vegetables in plastic bags to prevent moisture loss, but this can be tricky. The bags should be perforated to permit ventilation, or surface molds that thrive on stagnant air will take over and spoil the vegetables.

It's important to remember that cool air can absorb less moisture than warm air. For this reason, you get a pretty unstable situation in a place that's both cold and very damp. A slight drop in temperature can mean that air that, at 34 degrees F, let's say, had room for a bit more moisture, suddenly becomes fully saturated at 32 degrees. When this happens, you've reached the dew point, and the excess water that the air can no longer hold begins to condense on wall, ceiling, and even vegetable surfaces. When produce becomes wet it's more likely to spoil, even at low temperatures. So you see that your aim is to achieve a balance.

Since it is impossible to control all these factors entirely, you will often have some condensation, especially in dug-in underground root cellars. Therefore it is a good idea to keep the ceiling surface smooth, with no beams or other structural parts projecting into the room. When a beam protrudes, it traps warm damp air in the poorly ventilated places where it joins the walls and ceilings, and this causes undue condensation. This is why some old, carefully built root cellars have arched ceilings, with no corners to trap warm air. If condensation is a problem in your cellar, you might want to spread dry burlap sacks or layers of newspapers over the food to absorb any water that might drip off the ceiling.

## Ventilation

Ventilation, the third key to successful root cellar operation, affects both temperature and humidity. Admitting cool night air to the cellar, as we have seen, helps to chill the area to the desired low temperature more quickly in the fall. Equally important, adjusting air intake can help to reduce excessive humidity and thus prevent undesirable condensation. In addition, the air circulation set in motion by a good ventilating system effectively removes both vegetable odors and ethylene gas given off by stored fruits, which might otherwise produce off-flavors in other foods and sprouting in potatoes in stagnant air.

To understand how air can circulate through a root cellar, remember that warm air rises and cool air falls. In a large or tightly enclosed cellar, you'll need both an air intake and an air outlet. The intake opening should be low, and the outlet should be placed high and ideally

## Air Flow Patterns

Figure 1

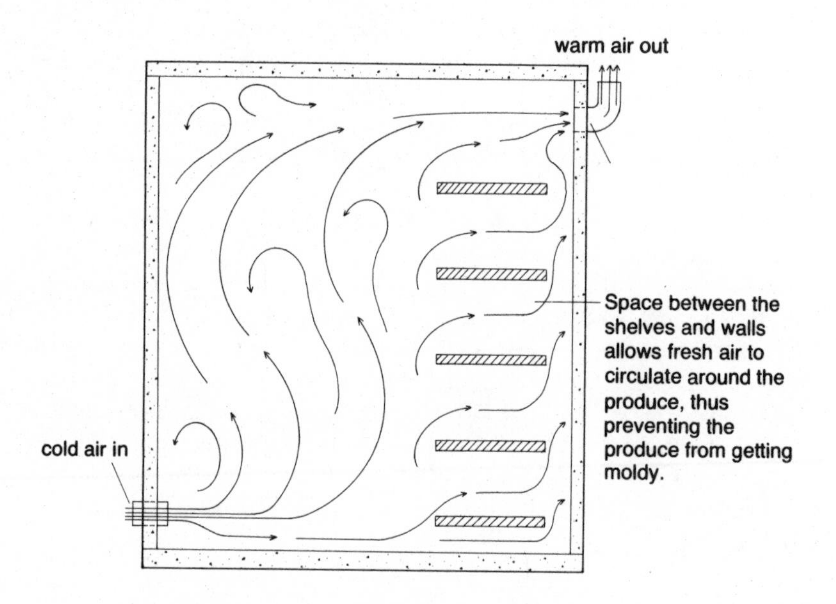

## Air Flow Patterns

Figure 2

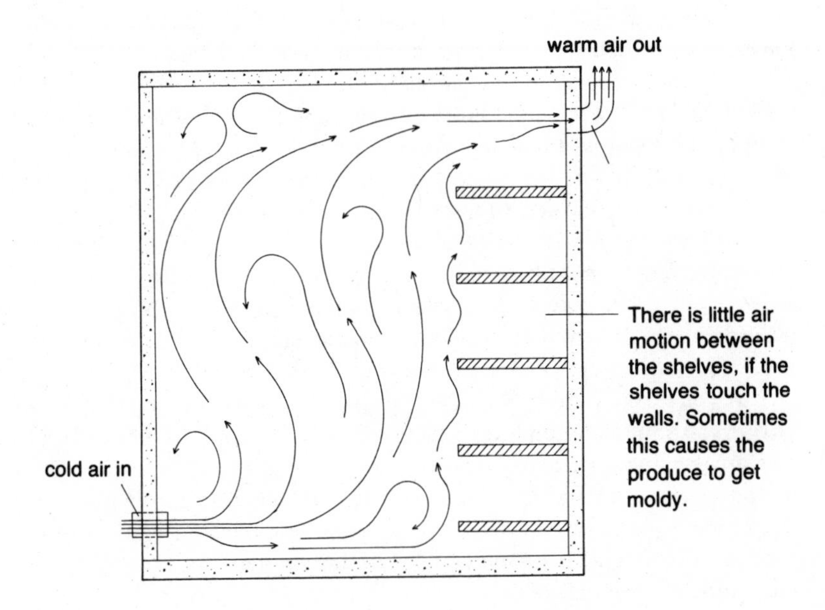


Figure 3

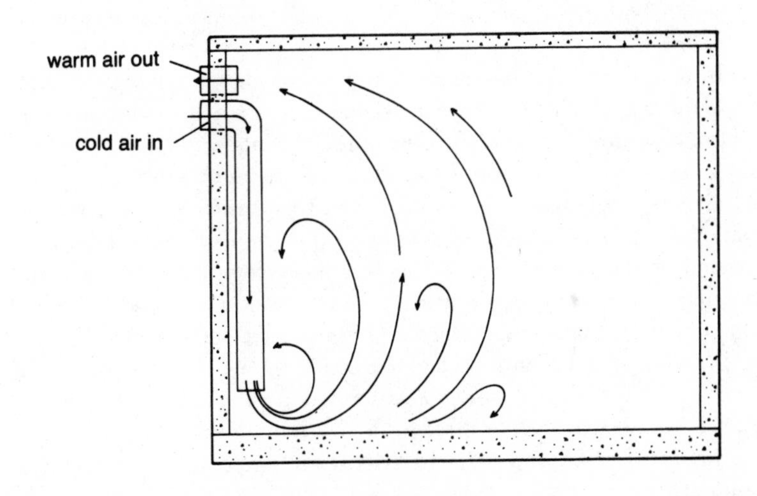

on the opposite side of the room to promote more complete circulation of air. Cool air will enter through the low intake and warm air will be released through the outlet. In a small storage area or one with many cracks where air can enter, such as an old stone foundation, a single outlet, placed high to exhaust warm air, may be sufficient.

The diagrams show the air flow pattern in three different root cellars. Figure #1 with the air intake and air outlet vents on opposite sides of the room and a space between the shelves and the wall, is an ideal setup. In figure #2, the cellar is the same except that there is no gap between the shelves and the wall for air circulation. Figure #3 shows a large cellar with intake and outlet vents on the same wall. Such an arrangement would be adequate for a small space — five by five or so — but in a longer room the incoming cold air is blocked by a curtain of rising warm air, so it tends to stay near the entrance. (As the cold air moves into the room, it warms up and then rises.)

If you're building a new concrete block or stone root cellar, we'd advise you to include both air intake and exhaust pipes in your plans. Use any reliable plastic or metal drainage pipe, clay tile or scrap pipe, or commercial ventilating blocks, and screen the exterior openings to keep rodents out. For root cellars six by eight feet or smaller, use a vent pipe with a four-inch diameter. Cellars larger than six by eight will often function better with a six-inch pipe.

When installing the exhaust pipe, keep its open edge flush with the cellar wall. It should not protrude inside the cellar, because the dead air pocket thus formed would encourage condensation. In some European root cellars both the intake and exhaust pipes are insulated to prevent condensation.

In a well-planned basement storage room designed by the Canadian Department of Agriculture, cross ventilation is achieved in the area by running a fresh-air duct from the outdoor window along the wall for several feet, ending with a damper that can be flapped shut to exclude cold air when necessary. Warm air exhaust in this Canadian system is handled by a small fan placed in the other half of the window not occupied by the intake duct. To operate this system automatically, the Canadian specialists suggest using a differential thermostat with one bulb kept inside the storage space and the other placed outdoors where the sun will not shine on it. When the inside temperature rises above a preset level, then, the exhaust fan will automatically start to vent warm air and cold air will rush in through the open intake duct to replace the outgoing air. The fan shuts off when the room temperature is reduced to the point you've chosen.

## Installing Vents Properly

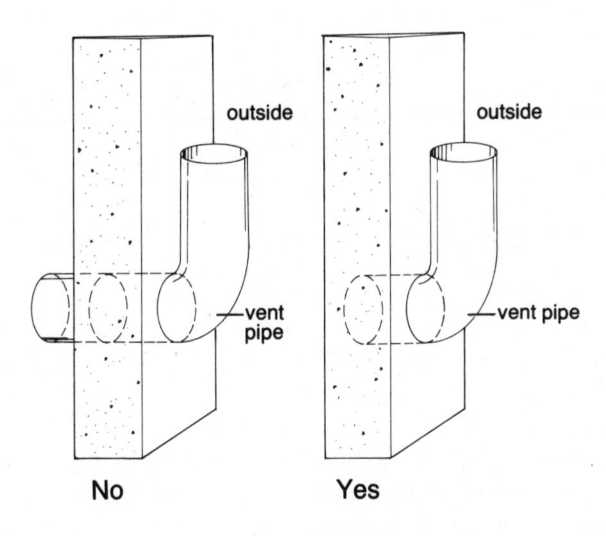


If extra-cold temperatures threaten your stored food, you might wish to use the simpler plan of installing a thermostat in the root cellar with a bell, light, or other indicator, outside the cellar, set to go off if temperatures drop dangerously low.

Here again, these are refinements that might be considered if your area experiences severe freezes or if your stored food supply is precarious or vital enough to make them worthwhile. None of the root cellaring folks we interviewed had found it necessary to use any of these mechanical aids.

After providing for the three factors vital to a workable storage area — temperature, humidity, and ventilation, you'll want to make note of some other considerations as you plan for your vegetable independence system.

## Accessibility

The more convenient your root cellar is, the more useful it will be to you. As we mentioned above, though, your average winter snowfall and amount of drifting will influence the placement of your cellar. In parts of Canada where snow is deep, it's the custom to build root cellars under the living quarters or an attached porch, often with access by a trapdoor and ladder. People would rather negotiate the ladder with an apron full of apples than slog through three feet of snow in a biting wind to an outdoor root cellar. Other things being equal, though, a root cellar that is close to the house, easy to clean, well lit (a bare light bulb is fine), and not too unhandy to climb into will give you the most pleasure and satisfaction. When possible, have the door open outward in case the floor is jammed with bins and crates. A well-built root cellar lasts a long, long time so extra care in planning for convenient use is worth while.

## Darkness

It's true that an electric light (or an available kerosene lantern) is a big help when you go to plunder your store, but between times the root cellar should be dark. Light deteriorates some storage vegetables and encourages sprouting in potatoes. Sun shining into the room can also raise the temperature higher than it should be. If you ventilate a basement room by raising an outside window, be sure to shade the window.

## Drainage

You want your root cellar to be damp, it is true, but you don't want it to be waterlogged. The food should never be in direct contact with water. It's a good idea, especially in clay soil, to lay a three-inch bed of 2B-size gravel around the perimeter of an underground cellar and put perforated drainage pipe on the gravel to carry off excess water. We've seen some good old basement root cellars in which humidity was kept high by running water, but that water was an underground spring that was *channeled.* It entered through a pipe, coursed through a concrete trough along one wall and ran out the low end of the cellar through another pipe. Few partitioned cellars have a problem with drainage; usually, especially if the floor is concrete, they'll need added humidity.

## Materials and Shelving

Use nontoxic materials in this food storage area. We wouldn't recommend creosoted wood, for example, for bins or shelves that might touch the vegetables. When ventilation is poor and humidity high, shelves made of pine or spruce may rot within several years, especially if they are not slatted to allow for drainage of condensed moisture. If you have your own woods, you might find it worthwhile to cut cedar, oak, or locust for lumber as we have done. Locust logs should be seasoned for a year before being milled into boards. We learned, when we took fresh locust logs to the sawyer, that it is nearly impossible to saw boards of even thickness from freshly cut locust. The slightly wedge-shaped boards we got from our green locust logs were fine for rough construction, but the logs would have yielded more boards, and straighter ones, if they had been aged.

Oak is so strong and durable that you can use one-inch oak planks for shelving. Next best would be Douglas fir, but you'll need to use two-inch boards for shelves made of fir. Pine and spruce, less durable woods, should also be two inches thick for shelves. With good ventilation, untreated pine boards should last for five to eight years. Make all bins and walkways slatted, to encourage air circulation. If you can make shelves movable and bins small enough to lift easily, the necessary annual cleaning of the root cellar will be easier.

## Crates vs. Half-Bushel Baskets

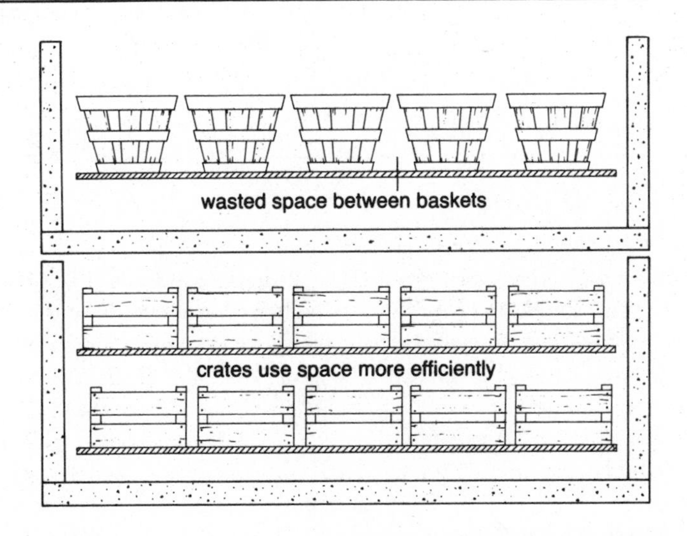

Half-bushel baskets are light in weight and easy to handle. Crates utilize space more efficiently and they may be stacked. Here is a plan for a sturdy storage tray that stacks well and permits good ventilation. Convenient dimensions would be about 16 by 22 inches and 3 to 12

## Storage Crate

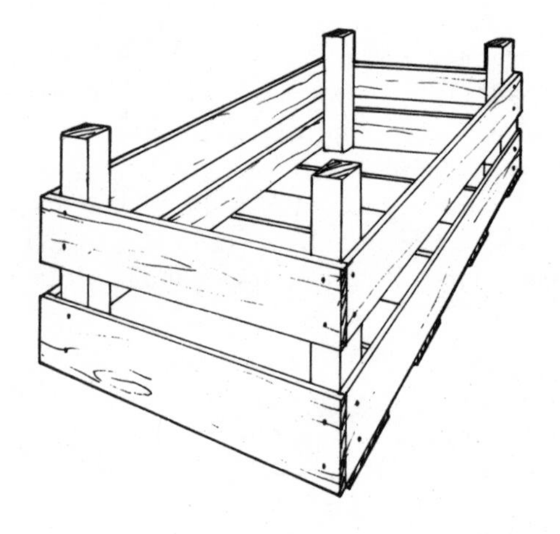

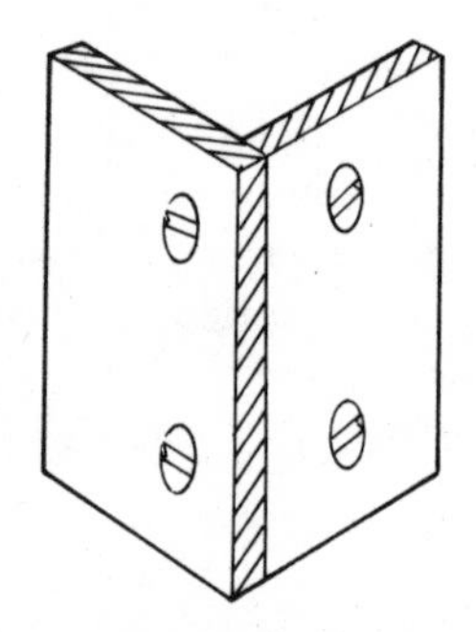

reinforcement for corner made of 2 small pieces of wood

inches deep (3 inches for grapes, 6 to 12 inches for apples or potatoes). Corner posts should be 2 to 3 inches higher than the body of the crate.

Nail two slats on each side to corner support posts. Then nail ³⁄₈-inch boards across the bottom. The posts hold the crates apart when they're stacked. For extra heavy use add reinforcements at the corners.

The use of salvaged materials can often reduce construction costs considerably. Keep track of what goes into your root cellar and how much you spend. Within five years you may well consider this one of your better investments.

## Size

How large an area do you need? Not as large as you might think. You can fit a lot of produce in a small area. A room measuring five by eight feet, for example, can hold 30 bushels of produce. If you intend to keep tubs of green vegetables with their roots still on in soil or sand, you'll need somewhat more room. Most of the formal plans for root cellars published by state agriculture departments recommend enclosing a space ten feet square. In our consultations with root cellaring gardeners, we noted that not all of the cellars we visited were filled to capacity with vegetables and fruits. If you have the room, though, and want to make good use of the backhoe you've hired to dig out the space, you might want to make your root cellar a bit on the large side rather than too small, with an eye toward future expansion of your gardening activities. A space eight by eight feet should be plenty for the average family, and ten by ten feet should take care of all you can manage to produce.

Several of the gardeners we visited shared their root cellars with other members of their family. If you live in a community of gardeners or homesteaders and have an especially good site for a root cellar, you might want to consider setting up a cooperative root cellar in which storage facilities and maintenance tasks would be shared by several families. Building the cellar could be a cooperative venture too. Or an individual owner could charge a small rent or accept produce or other bartered goods for a season's storage space in a large-capacity privately owned root cellar. As with any cooperative enterprise, it's a good idea to spell out all terms clearly before starting.

## Canned Goods

My Aunt Margie, whose gift for vivid description is legendary in our family, tells a marvelous tale about the day her shelves full of canning jars — dozens of them, all full, of course — collapsed noisily and messily all over her cellar floor. Sad indeed. That's only one of the things that can happen if heavy jars are kept on wood of uncertain strength in a damp place.

More common, as we've discovered in the damp dirt-floored basement where we kept both root vegetables and canned goods before building our new house, is that the jar lids rust through in several years. Any canned goods you intend to keep longer than a year in a damp place should be checked before using, because as the rust invades the metal lid, it eventually loosens the seal, allowing air to enter the jar. In our case it took about three years for the lids to rust enough to loosen them. If your cellar is humid enough for root vegetable storage, it is too wet for canned goods. They need a *dry,* cool spot. We've solved that problem in our new root cellar by partitioning the damp earth-floored root cellar off from the slightly warmer and considerably drier surrounding basement. If you're starting at the beginning, then, plan a separate place for your canned goods. If it's dry enough there, you can store your onions or squash there too, depending on the temperature range.

# 14

# Keeping Things Humming in the Root Cellar

In leading kolkhozes (Russian collective farms) the wages of workers responsible for storing and of the storemen are graded according to the storage results. Workers who succeed in reducing losses below those of standard get a bonus according to the value of the additionally marketed produce.

E. P. Shirakov,
*Practical Course in Storage and Processing of Fruits and Vegetables*

A root cellar needs a certain amount of maintenance in order to keep it running well. If your root cellar is well stocked, you'll probably be visiting it nearly every day, anyway, so you can easily keep an eye on things.

Watch the temperature. If the thermometer had registered 32 degrees F, but has now risen to 38 or 40 degrees, open the cold-air vent to cool the room. You may need to fool with the air vents every day or so to keep the place cold but not too cold. The idea is to ventilate the cellar well enough to carry away heat and ethylene gas given off by the vegetables and fruits as they breathe, but not to set up such a strong draft that the rapidly moving air removes moisture from the produce. So experiment with different ventilation settings and observe what happens. Keep the intake vent closed during the day if the outdoor temperature is higher than that in your root cellar. In such cases, the best plan is to blow cold night air into the cellar with a fan.

When the outdoor temperature is lower than 15 degrees F, it doesn't take long to chill the cellar by admitting outdoor air. In fact, in times of bitter cold, below-zero weather, you'll want to take care, when entering an outdoor root cellar, not to admit too much of that frigid air. Double doors help here. And in warm weather, close the door promptly when you enter. Frequent opening of the root cellar door on a warm day can raise the temperature of the storage space to undesirably high levels.

During the coldest part of winter, keep your thermometer in the coldest part of your root cellar so that you will know the lowest temperature in the cellar. In most indoor basement root cellars, winter temperatures will be lowest near the floor. In outdoor underground root cellars, especially those that are not deeply buried, the ceiling may be freezing while temperatures on lower shelves are just above 32 degrees F. Many root cellar owners aim for an average temperature of around 35 degrees so they don't need to worry about sudden dips in temperature freezing their produce.

If the outdoor temperature falls to minus 20 degrees F, even dug-in root cellars may be chilled below freezing inside. During such bitter cold weather some people leave a light bulb on or a kerosene lamp burning in the root cellar. (Cover the potatoes to protect them from the light.) When we have minus 5 degree F nights here in south central Pennsylvania, we fill one or two old buckets with hot coals from the wood box stove and set the buckets on metal trays on small mounds of sand, well away from paper stuff, in our cold dirt-floored basement.

Temperature is most difficult to regulate satisfactorily in early fall and again in spring when air is warm and many vegetables, especially cabbage, are at a stage in life where they produce more heat and release more moisture than they did earlier.

In areas with heavy snow, people sometimes transfer their stored cabbage to outside clamps in early spring just after the start of a thaw, when root cellar temperatures threaten to rise. They pile the heads in pyramids, packing two to four inches of snow between the heads and at least four inches over the top of the pile. Carrots, onions, and apples are sometimes put in boxes and packed in a mound of snow. Potatoes, also, may be piled on straw mats, covered with straw, protected by an additional layer of moisture-proof paper or plastic, and finally buried under a thick cap of more snow. Such a procedure is, of course, practical only in far northern areas where snow cover is heavy.

If humidity runs low (under 70 percent) or if the vegetables begin to shrivel, you can easily add more moisture by sprinkling water on the floor. In a finished basement storage room, spread damp towels or burlap bags over the produce and put shallow pans of water on the floor. Shallow containers are more effective humidifiers than deep buckets of water because the air currents in motion over their more accessible surfaces can carry off more moisture.

To raise humidity in a dry cellar you can also temporarily close the vents to decrease air circulation until you can get more moisture into the air.

To dry things up a bit when the cellar is too damp, open the vents to admit cold air and release warm air. Why? Cold air is relatively dry and when it mixes with the warmer air in the root cellar it warms up, and absorbs some of the excess moisture. The result is that the relative humidity of the air is decreased.

Another way to reduce moisture would be to put shallow containers of dry air-slaked lime or calcium chloride in the cellar. These materials absorb water. Remove and dry or exchange them when they become moist. They should not be put in contact with the vegetables.

Every week or so you'll want to examine your vegetables and fruits and discard any that might have spoiled. Often you can cook and use, or feed to animals, any produce that has begun to soften or shrivel.

Vegetables and fruits should always be handled carefully and kept in small lots to prevent bruising. You can minimize damage, too, by avoiding containers with splinters or large staples on the inside, which might puncture your foods. And even in a cellar that *seems* dry, keep crates of food a few inches off the floor so air can circulate around them. Slatted platforms under bins are simple to build. Pallets or sides from crates used to pack snowmobiles or industrial equipment also work well.

At the end of the vegetable storage season, it's a good idea to clean out the root cellar. Toss all spoiled vegetables on the compost pile. Clean out crates and baskets and expose them to hot sun for a day. Wipe down shelves and put them out to air if they're portable. And make some notes to yourself about what has gone well and what you want to improve in your root cellar operation, so you can learn from your experience and have an even better season next year.

# DO

1. Keep fruits and vegetables in small piles, not heaped in large mounds.
2. Handle produce carefully.
3. Check stored food often and weed out questionable specimens.
4. Store only your best—sound, unbruised, and mature.
5. Keep vegetables as cool as possible between harvest and storage.
6. Provide for ventilation in your storage place.
7. Use frozen onions and cabbage but don't let them thaw and refreeze.
8. Make use of leafy tops of vegetables when you harvest them.
9. Plant your root crops in deeply worked soil.
10. Cut leafy tops of root vegetables back to within an inch of the crown before packing them away.
11. Harvest storage produce in cold weather when soil is dry.
12. Keep a record or simple map of what you have stored and where, if you have produce scattered in sheds, garden, pits, and cellar.
13. Keep containers of vegetables raised several inches above the floor so air can circulate around them.
14. Pack root vegetables in sand or sawdust that's damp but not so soggy that it encourages rot or sprouting. If your cellar is dry, you may need to add a bit of moisture to the packing material once or twice during the winter.
15. Eat up what you've stored. It's there to be used.

# DON'T

1. Wash root vegetables before storing them.
2. Put vegetables right on a bare concrete floor.
3. Store insect-damaged, bruised, or immature produce.
4. Keep onions, garlic, squash, pumpkins, or sweet potatoes in a damp place.
5. Store stemless squash or pumpkins.
6. Give root crops or storage fruits large doses of high-nitrogen fertilizer.
7. Seal incompletely dried nuts and grains in tightly closed containers.
8. Feel badly if a small percentage of your stored produce spoils. You're still way ahead growing and keeping your own good food.

# 15

# The Basement Root Cellar

I dug my cellar in the side of a hill sloping to the south, where a woodchuck had formerly dug his burrow, down through sumac and blackberry roots and the lowest stain of vegetation, six feet square by seven feet deep, to a fine sand where potatoes would not freeze in any weather. . . . I took particular pleasure in this breaking of ground, for in almost all latitudes men dig into the earth for an equable temperature. Under the most splendid house in the city is still to be found the cellar where they store their roots as of old, and long after the superstructure had disappeared posterity remark its dent in the earth. The house is still but a sort of porch at the entrance of a burrow.

Henry David Thoreau
*Walden*

Basement root cellars are convenient to use and relatively inexpensive to build. Your modest investment in lumber and insulation will make it possible for you to store more good home-grown food for winter without the bother of canning and freezing. You might even find it practical, if your garden is small, to grow some additional produce in a community garden. You could, for example, use your community garden patch to raise a whole plot of good keepers — squash, onions, leeks, cabbage, carrots, beets, and such, especially for winter storage, while you grow the perishables like peas and lettuce and snap beans in your backyard garden.

If you live in an old house, you may even already have a dirt-floored room or corner in the basement. The original builders set this area aside especially for vegetable storage. After all, they didn't have freezers, and they often had large families to feed. You can easily reclaim such a cellar. If it has a window, you can readily cool and ventilate the room. If a furnace in the cellar heats up the dirt-floored room too much, you can tack insulation around the walls of the room and around any hot pipes or ducts that may go through the room. Sometimes an old coal bin can be turned into a root cellar. It is best if the room is on the northeast side of the house, or, second best, the northwest side, because these are the coldest corners.

The cellar of an old house that's heated by wood rather than a furnace is even easier to convert to a root cellar. Sometimes you can use

The dirt-floored root cellar in our old house.

it just as it is. In the old house here on our farm, the whole floor of the 30- by 35-foot cellar is dirt. There's a modern oil furnace in the cellar but we don't use it. We heat the house with a wood box stove. A gas hot water heater in the cellar adds a small amount of heat, but the area remains cold and damp.

When we first moved here to the farm, the cellar ceiling was so low that we could not quite stand up in it. That first spring, Mike laboriously hand-dug a foot of soil and cinders from the floor, hacked out a trench around the perimeter of the old stone walls to channel excess rainwater that sometimes entered the cellar, put a layer of six-mil plastic down over the floor and covered that with two inches of sand.

Thanks to wood heat and Mike's engineering, our whole old basement is now a root cellar and we keep all kinds of root vegetables, green leafy vegetables, apples, and citrus fruits down there with very good results.

There are probably many more gardeners, though, who have finished basements in which a furnace keeps the basement area pretty warm — too warm, at any rate, for vegetable storage. If this is your situation, consider partitioning off one corner on the north side of the basement to enclose a food cellar. Try to include one outside window or, if this is not possible, arrange for vent pipes or a screened vent block (sold by lumberyards) to be inserted through the foundation.

Unlike an outdoor excavation, this job may be done on rainy weekends, dark evenings, and chill winter days. If you've done any house or shed construction, you'll find this a simple project. For many people, though, enclosing a room will be a new experience — one that may seem confusing and a little bit overwhelming until you get the hang of it. Don't worry. You can do it. When we moved to our farm six years ago the only thing we'd ever built was a small three-sided goat shelter. Since then we've put up a barn, a log shed, a garage and — most recently — the shell of a new house. In the process, we've discovered that even a big job becomes manageable when you break it down into parts and proceed, without hurrying, one step at a time.

So let's pace off your warm, furnace-inhabited basement and lay out a root cellar for you, one step at a time. Work on it gradually, measure carefully, keep hoeing that potato patch, and before the year's out your house will have a whole new dimension and you'll be several notches more independent of grocery store food and outside energy sources.

Mike's first root cellar plan encloses a small space, only 3½ by 7 feet. That's about the minimum for a room you can enter, and it doesn't

## Basement Root Cellar Number 1

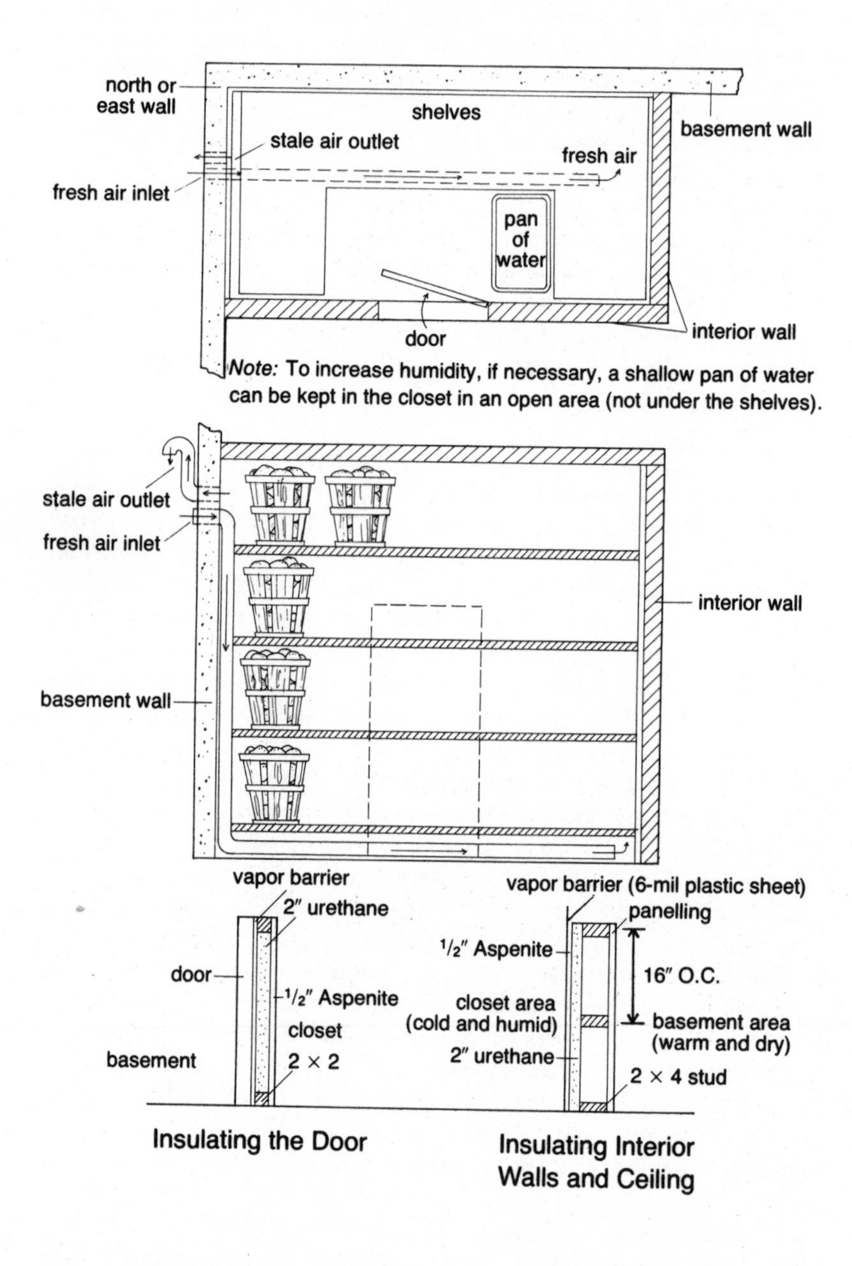

really leave much space for closing the door behind you. This small space can hold 28 half-bushel baskets of produce, though. If you have only a limited area to devote to food storage, you could follow this plan and add a floor-to-ceiling curtain on the long exposed side of the cellar to keep out warm air when the door is opened. Making the storeroom deeper or wider would provide enough room to close the door behind you when you enter the cellar.

Or, if you can't spare even this much space, you could build a vegetable hideaway closet rather than a room simply by putting up shelves, enclosing them with studs and insulation, and putting two insulated doors on the front for easy access to all corners of the root closet.

We've suggested using Aspenite to line the closet because we have found this material to be durable in a damp place, yet it is less expensive than exterior plywood. (It is composed of wood chips pressed together with exterior glue.) For a less expensive liner you could use aluminum offset press plates which are often available from newspapers for less than a dollar apiece (often much less). These are more easily dented so it would be necessary to use care in putting baskets close to the walls. Imaginative recyclers will find other lining materials to use — such as old linoleum rugs, scraps of Formica, and panelling sometimes available at dump and construction sites.

**Step One** Attach a sill of two-by-four-inch lumber to the floor. Mike would use pressure-treated lumber for this sill if the floor is likely to be damp, but regular lumber should serve the purpose very well as long as the vapor barrier you'll put on later goes clear down to the floor, to keep the dampness in the root cellar away from the wood framing.

To fasten the wood two-by-four to a concrete floor, first drill holes through the wood every two to three feet, using a wood drill bit. Then place the sill in position on the floor and, using the predrilled holes as a template, drill holes in the concrete with a masonry drill. Sweep away the fine concrete dust. Run a bead of construction cement around the edges of the two-by-four. Replace the two-by-four over the predrilled holes. Insert sleeves and then screw in lag bolts, taking care not to strip the threads. Your sill should run around the perimeter of the root cellar, leaving a 30-inch rough opening for the door.

**Step Two** Toenail two-by-four-inch studs to the plate every 16 inches on center.

**Step Three** Nail a two-by-four-inch header on top of the studs.

**Step Four** Fasten a vapor barrier — tar paper or six-mil polyethylene sheeting — to the studs. The vapor barrier should always be put on the side of the insulation facing the warmer area.

**Step Five** Attach insulation to the studs. (Some Styrofoam and fiberglass insulation comes with vapor barrier already attached.) Styrofoam or urethane panels are less likely than fiberglass to become waterlogged in a very damp cellar.

**Step Six** Hang an insulated door in the door frame. Use a standard door with two-by-two-inch lumber nailed all around the edge on the inside. Fit two-inch Styrofoam inside this frame of two-by-twos. Cover the Styrofoam with Aspenite or other improvised panelling.

**Step Seven** Install the vent pipes. If buying new ones, we'd use plastic pipe because it won't rust, but make use of what you have. There's a reason for carrying the intake pipe across the wall to the opposite side of the room. The only way to get air circulation is to make the air *travel.* If the intake pipe is too close to the exhaust pipe you have almost a closed loop and the air doesn't go anywhere in the cellar. You'll also notice that the exhaust pipe is carried *above* the outlet level before opening into daylight. There's a reason for that too. Making the opening of the pipe outlet at the same level as the horizontal section of the pipe would retain a blanket of stagnant air at the top of the root cellar. With a higher outlet opening, stale air is forced up and out and even the uppermost few inches of the cellar receive fresh moving air.

The fact that the wall is wet from condensation does not necessarily mean that humidity in the cellar is high enough. Warm air in contact with the cold wall loses moisture, which shows up as droplets of condensation. If you live in a very cold place where the temperature of the outside wall is quite low, you may want to insulate the upper 12 to 24 inches of the wall to prevent loss of air humidity by condensation on the wall.

If the door does not fit as tightly as it should at the bottom, use weather stripping and shove an old rug or coat up tight against the crack to prevent warm basement air from entering the root cellar.

The second basement-access root cellar is one Mike devised based on a similar cellar we visited, which had no ventilation or insulation. The owner built this food cellar when he built his house. Access to the cellar is by way of a door in the interior wall of the garage, which was situated under the house. A concrete patio tops the root cellar.

The third root cellar plan is for a larger partitioned basement root cellar. The rectangular space is efficient, and allows you to position one long wall along the exterior cold foundation wall for more effective cooling.

## Basement Root Cellar Number 2

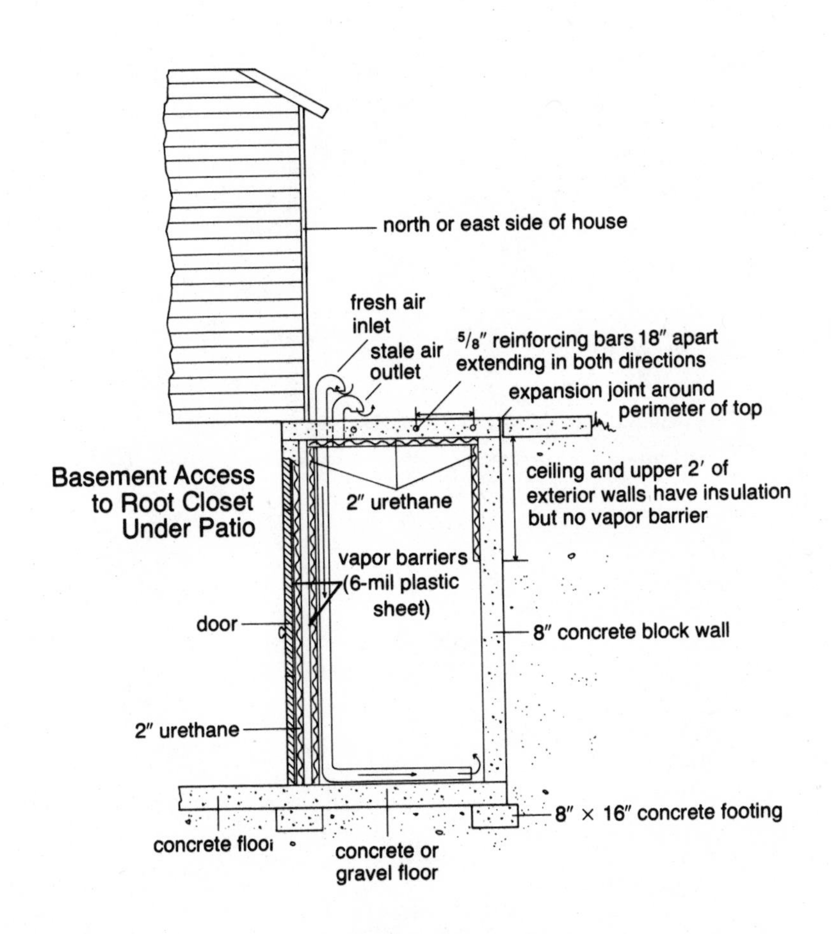

## Basement Root Cellar Number 3

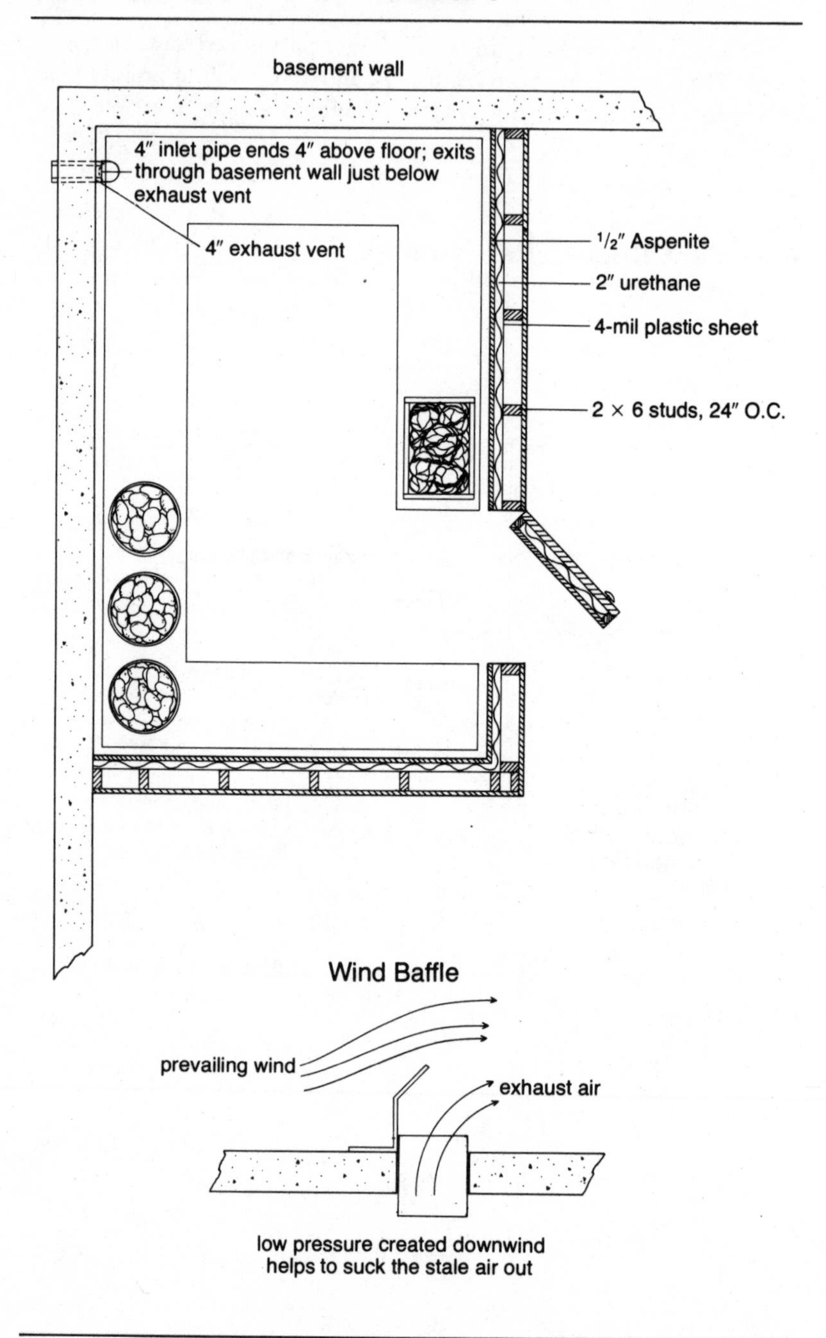

## Building Shelves

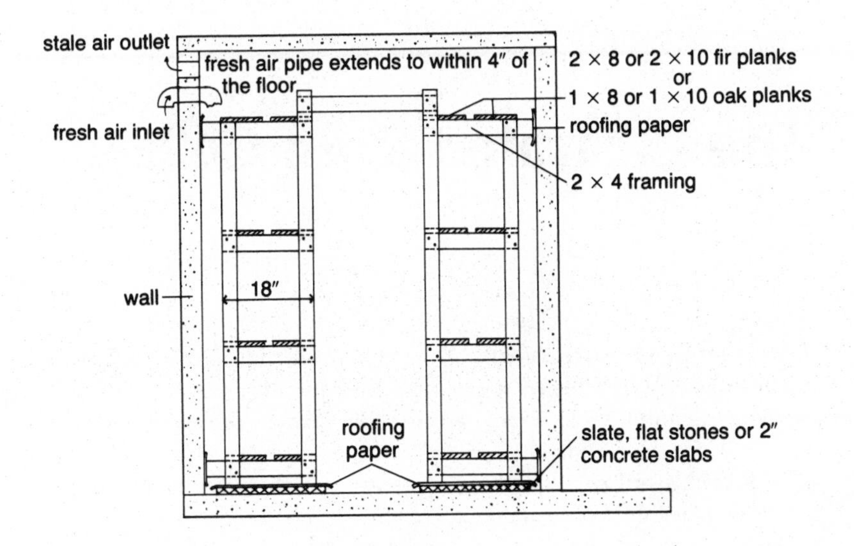


Side View

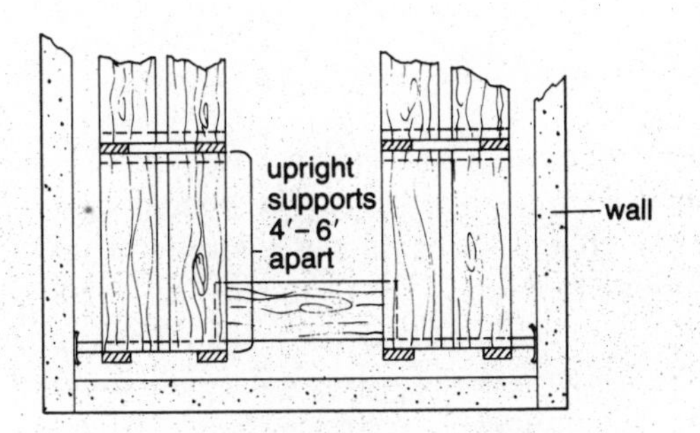


Top View

*Note:* The upper and lower crosspieces are longer than intermediate ones to keep shelf units away from the wall for good ventilation.

# 16

# The Excavated Root Cellar

The cellar in (the early nineteenth century) was not always under the house. It was more often off to one side. Usually it was on the north side of the house, and it was merely a room dug deep in the ground, with a dirt floor, for the storing of foods. *Cellar* was actually a mispronunciation of a French word *cella,* meaning *store-room* . . . the word cellar meant *pantry* or place for food storage.

Eric Sloane
*Diary of an Early American Boy*

Dug-in root cellars work well because they are insulated by the earth surrounding them. The soil is a poor conductor of heat, so the temperature of the ground six feet under the surface is cool and fairly constant. The natural moisture of the earth helps to keep humidity high. Soil is heavy, so an underground root cellar should be strongly built. In addition, because frozen wet soil can expand and rupture walls, it is important to provide drainage around the cellar so there is no water-logged soil to freeze and cave in the walls of your cellar.

Mike has designed several classic root cellars with these principles in mind. Observation of root cellars other folks had built and experience in doing all the construction work on our own house helped him to plan his versions of the ideal root cellar. These plans for in-ground root cellars, as sketched in the following pages, will show you what is important to include, so you can adapt the designs to your own situation.

# Dug-In Root Cellar Number 1

The two rooms in this root cellar have different temperatures and humidities.

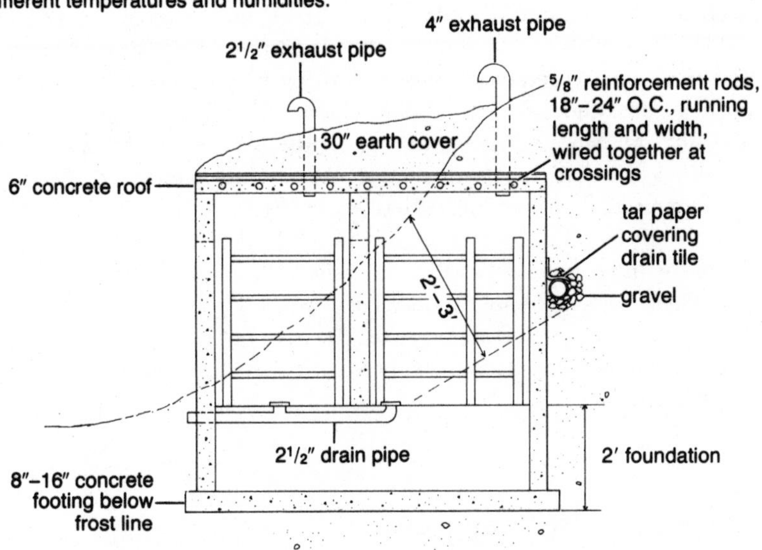


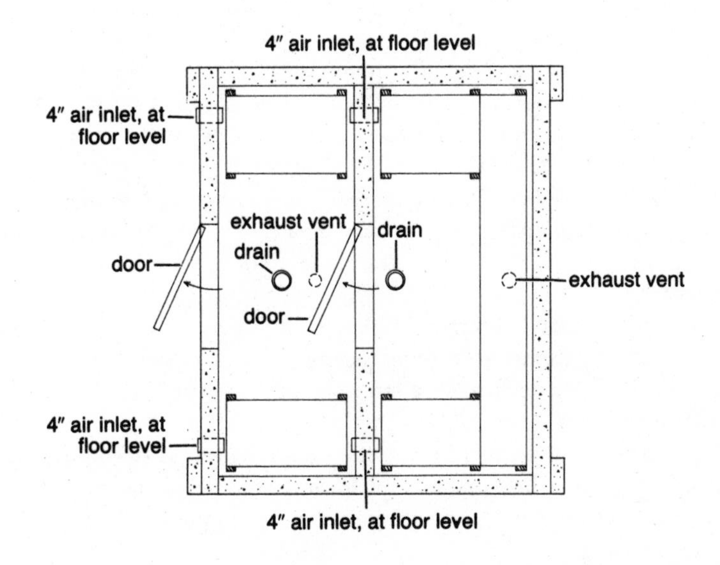

You'll notice that both of the cellars are divided into two parts. This division serves several purposes. The small anteroom helps to insulate the inner chamber from extremes of heat and cold. It also provides a choice of conditions. In winter the outer room is colder. In severely cold weather it may go below freezing. In summer it's warmer. You can fine-tune your vegetable storage system by moving produce from one room to the other when conditions change. In a large cellar, the wall dividing the two areas helps to support the roof. A cellar with the dimensions sketched here will hold 44 half-bushel baskets in the large inner room and 16 half-bushel baskets in the small outer room, which, however, is useful for storage mainly in the spring and fall. (The capacity of the inner room is based on four shelves, each holding 11 baskets.)

Construction of either root cellar means working with concrete and laying up concrete blocks. For those who may not have experience with concrete or concrete block work, here's a quick mini-course to help you get started.

## Working with Concrete

The best temperature for working with concrete is that at which you yourself are most comfortable — 60 to 80 degrees F. On very hot days concrete cures more rapidly than it should. Concrete is stronger if it cures (dries and hardens) slowly. For that reason, it's a good idea to plan your pour for moderate weather and wet down the concrete or cover it with damp burlap for the first 24 hours. Fresh concrete shouldn't be allowed to freeze either. If you must pour late in the fall, cover the work to help it retain the heat generated by internal hydration — the process by which the water in the mix bonds chemically with the dry matter.

Concrete poured directly on the ground should be put on firm, undisturbed soil. Remove all topsoil before pouring. Recently filled land that is likely to settle further is not a good place to pour a concrete foundation or floor.

When pouring a concrete foundation you can add strength to the structure and save on concrete by tossing rocks into the foundation trench. All rock fill should, of course, remain below the projected surface level of the concrete.

Concrete has lots of compressive strength — when it's well supported, you can put all kinds of heavy weight on it. But when poured concrete must span a roof or other unsupported gap, it needs some help. To strengthen such clear spans of concrete, embed steel reinforcing bars or wire mesh in the wet concrete (or use old pipes, bedsprings, or even rusty barbed wire).

Recruit at least one helper for your concrete work. For a big job, call out the whole family or several strong neighbors. Wet concrete ruins shoes and irritates skin, so wear gloves and rubber boots.

Rake, tamp, and jiggle the wet concrete as it's being poured to be sure that all corners are covered and no air pockets are trapped. Then screed (level) the surface with a straight two-by-four. Use a float (a rounded-edged piece of sturdy wood with a handle on one long surface) to press the concrete surface gently so that all stones and gravel are submerged in the liquid mixture. Finally, trowel the surface smooth when it becomes firm.

## Concrete Mixes

For small jobs you may want to mix the concrete yourself in a wheelbarrow or trough. You can rent electric or gas-powered cement mixers, but we find it easier to hand-mix small batches and order large batches already mixed from a transit-mix or metered mix-on-the-spot commercial concrete dealer. It's much easier to clean out a wheelbarrow or garden cart than a cement mixer. If you are doing your own mixing, thoroughly blend the dry ingredients before adding water. Use clean, fresh water. Gravel should not contain too much dust or silt. Add water slowly and mix thoroughly as you go. Clean out all containers and hose off shovels and tools as soon as you finish the job. Liquid concrete is not difficult to remove, but if it hardens on your shovel, that's permanent!

If you order from a dealer you'll need to know how much to get. To estimate how many cubic feet or yards of concrete you need, multiply the length and width of the area you want to fill. If your footing measures 2 feet wide and 10 feet long, that's 20 square feet. Then multiply that figure by the depth of the concrete in feet or fractions of a foot. A footing 15 inches deep would be 1¼ feet; thus $20 \times 1\frac{1}{4} = 25$ cubic feet. It's always a good idea to order a bit extra and be prepared to use any concrete left over to patch a walk or set a post. Running out of concrete in mid-pour is *much* worse than buying a bit too much! If you need a small amount of concrete, your best bet might be to order from one of the metered trucks that delivers exact quantities, even as little as one cubic yard. The price per cubic yard is usually somewhat higher than for transit mix, but you only pay for what you use, not for a whole truck load.

For a strong concrete mixture, use 1 part cement, 2¼ parts sand, and 3 parts gravel. A somewhat weaker mixture suitable for foundations and other protected areas may be made by mixing 1 part cement, 2¾ parts sand, and 4 parts gravel. Add just enough water to make the mix workable. Too soupy a mix makes a weak concrete.

### Forms

You need forms to hold the concrete in place until it dries. Forms may be made of scrap lumber, but they should be levelled and plumbed with great care and strongly supported. Wet concrete exerts enormous outward pressure. You can reuse lumber from forms if you brush it with used crankcase oil or cheap motor oil before it touches wet cement. Use double-headed nails to fasten forms together. They save time, temper, and equipment because they make the forms easy to disassemble. Support low forms like those used for footings by driving in pointed stakes. You may also need to nail struts between the form boards to keep them from spreading. Such struts are necessary when pouring a concrete wall. Internal braces — two-by-fours holding the side wall forms apart — are extremely important for poured walls. The forms for a poured wall are usually made of exterior plywood if they will be reused for another job, but interior plywood is adequate for a one-time job. These forms should also be braced by nailed-on two-by-fours, nailed on edge around the edges of the plywood. Have some extra braces close at hand in case you need to grab for them quickly while pouring. As the experts say, concrete doesn't wait!

## Concrete Block Work

Plan your building in multiples of 8 inches (half the length of a standard 16-by-16-by-8-inch concrete block including the mortar allowance). Mark the corners exactly with crossed strings tied to batter boards.

Concrete blocks should always rest on a poured concrete footing so the weight of the structure is distributed over a wider area of ground. Never set mortared blocks directly on the soil. Begin laying blocks at the

## String Layout

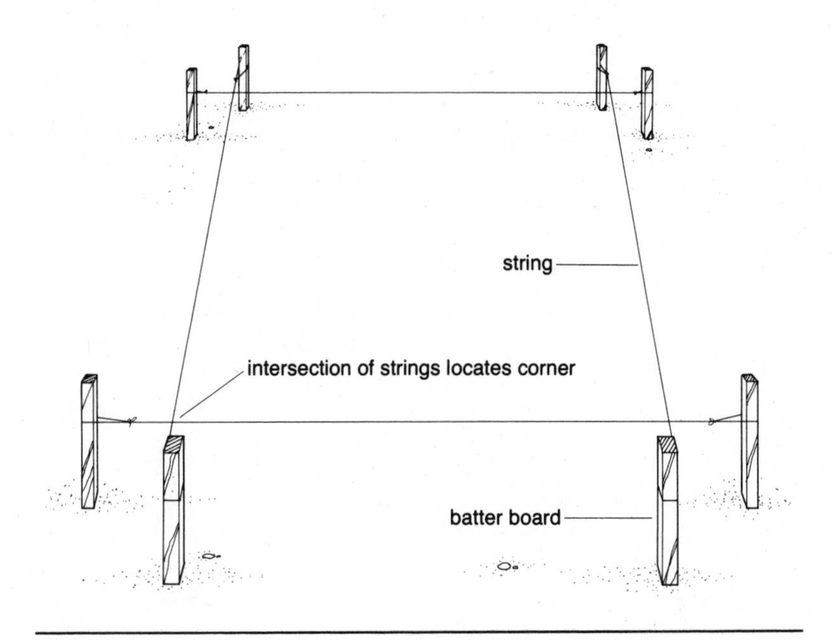


corners and build up several courses at the corners before filling in the center span of the wall. Level and plumb each course as you go. Don't shift blocks after the mortar has started to set.

Make mortar in small batches. When it dries and gets crumbly an hour or so after mixing it's not as good. Mortar over two hours old should be discarded. Use the same unit of measure for each batch you mix so that the mortar will be as uniform as possible. Good mortar holds its shape somewhat when it is dumped from a bucket but it glistens smoothly when trowelled. Use a mortar board — a foot-square (or more) piece of scrap plywood — to carry mortar where you need it. A mixture of one part Portland cement, one part lime, and four parts sand, blended with just enough water to make a plastic mix, makes a good mortar.

On inside or exposed walls, smooth and press the mortar joint with a jointing tool or a piece of scrap metal rod when the mortar is partly set but still plastic.

For more complete information about working with concrete and concrete blocks, consult *Working with Concrete and Masonry* by Max Alth or *How to Work with Concrete and Masonry* by Darrell Huff.

## Construction Step-by-Step

If construction work is new to you, this is a good place to begin. You can bury most of your mistakes, as long as the structure is solid! Just proceed one step at a time, and you'll find you can do it. Let's say you're lucky enough to have a north-facing hill and you intend to make a dug-in root cellar in the slope of the hill. You're ready to begin, but how do you start? Follow us:

**Step One** As you've surely guessed, your first step is to dig a hole in the hill. You'll need to remove enough soil to make space for the volume of the cellar you plan to build. It's no help to dig a hole much larger than you really need, for all that space will have to be backfilled. If you're digging by hand, as a surprising number of root cellaring gardeners choose to do, you'll no doubt find it easy to avoid this pitfall.

Keep all that dirt handy. You'll need it later to fill in around the cellar. Dig deeply enough so that the footer will be below the depth of the average frost penetration in your area. Most of the cellar will be well protected by soil. Dig out an extra two-to-three-foot depth around the door where the foundation will be more exposed to the cold. Fill this hole with gravel up to the base of the footer.

**Step Two** Lay drain pipe from the floor drains in the cellar out to daylight. The floor inside the root cellar should be just a bit higher than the level of soil right outside the door so that water will drain off.

**Step Three** Pour the concrete footer. After the ground in the cellar site is levelled, make forms from 2-by-8-inch lumber. The footer should be 16 inches wide for a concrete block wall and 8 inches wide for a poured concrete wall 4 inches thick. (Or, in other words, twice as wide as the wall will be.) Technically, of course, the footer will be only 7¼ inches deep, since that is the actual dimension of a piece of 8-inch lumber these days. If you hand-dig the cellar you can just dig a trench for the footer, but if you have the work done by backhoe you'll need to make well-supported forms to hold the poured concrete footer. Put several reinforcing rods in the concrete poured over the dug-out spot at the door. No reinforcing is needed for the rest of the footer. Let the footer cure for three to seven days before proceeding with the walls.

**Step Four** Start laying concrete blocks. Since you've sensibly planned the dimensions of your cellar in multiples of 8 inches (half the length of a standard concrete block allowing for mortar joints), you should have no trouble here. Leave a gap for a standard door. For vent pipes in side walls, either leave a gap between blocks or chip one block to make room for the vent pipe. When you complete the last course of block your cellar should be 6½ to 8 feet deep.

**Step Five** Make forms to support the poured concrete roof. The roof should be reinforced with crossed, tied reinforcing bars as shown in the diagram. The bars should be embedded in the bottom half of the poured concrete roof. If they're too close to the surface they don't do much to strengthen the roof. The pour is made directly on plastic-covered ¾-inch plywood, which is supported by joists resting on beams which are held up by 4-by-4-inch posts. You'll also need a form running around the roof perimeter to hold the concrete ceiling until it dries and hardens. Make this form out of 2-by-6-inch lumber and attach supports — as shown in the diagram — at the corners and at least every 4 feet along the sides. Without these supports, the weight of the concrete is likely to bulge the forms and then drip through the gap. Why cover the

## Supporting Forms for a Concrete Roof

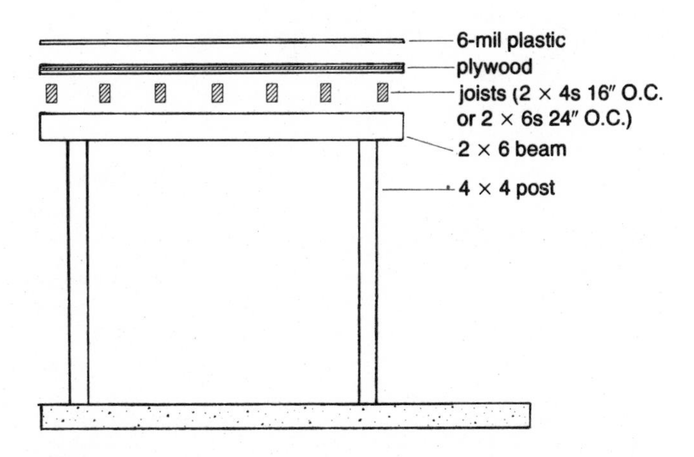

## Supporting Forms for Poured Concrete Roof

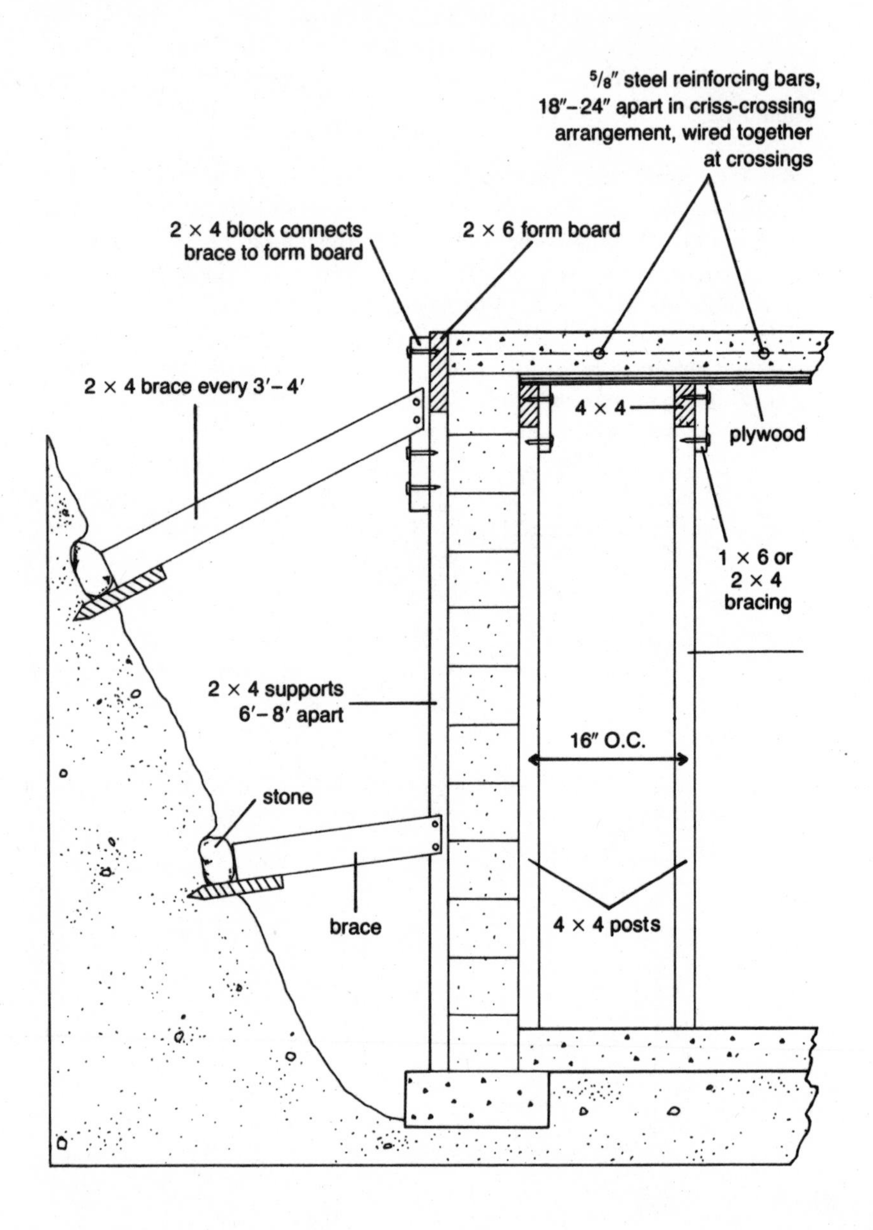

plywood with plastic? So that water from the concrete mix doesn't seep through and drain off. If that happened, it would weaken the concrete.

As you can see in the diagrams, your roof can be flat, peaked, or arched. The flat roof is easiest to build. The advantage of the peaked roof is that condensed moisture runs off more readily toward the corners rather than dripping directly on the vegetables. One shows an extra fillip — nice but not necessary — that you could include if you wanted to make a ribbed ceiling for even more effective channelling of ceiling condensation. On a ribbed ceiling the water droplets drip down the rib

## Root Cellar Ceilings

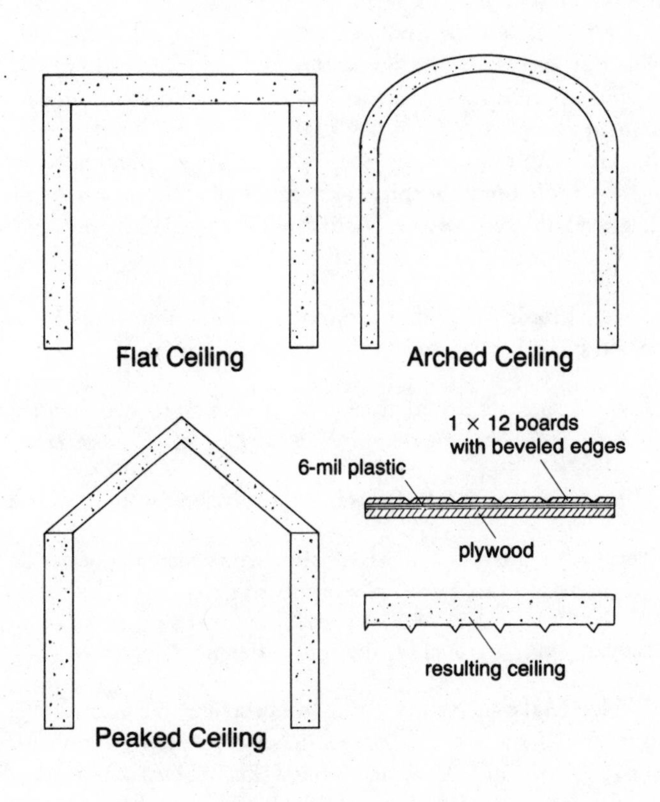

toward the side of the room. The arched ceiling has no dead air spaces and is easy to clean. To make it you can use heavy galvanized roofing rather than plywood as a base. It should be very well supported — wet concrete is heavy.

Remember to insert the exhaust and intake pipes in holes in the ceiling form board before you pour the concrete roof.

**Step Six** Start to backfill around the cellar. Tamp the soil lightly but firmly as you go. When the level of backfilled soil is still two to three feet below the projected soil surface height on the slope (see diagram), insert a perforated four-inch drainpipe running across the entire back of the cellar, bending around both corners and ending at daylight. The pipe should rest on a three-inch gravel bed and have a length of heavy roofing paper or six-mil plastic sheeting tucked partly over the top (but not around it) to prevent silt from clogging it. The pipe should slope one inch in eight feet starting from center back down both sides.

The purpose of this drainpipe is not to keep water out of the cellar but to prevent pressure on the rear wall of the root cellar from expansion of frozen wet soil.

Backfill with gravel above the pipe. If filling in manually, you can place a board behind the pipe and gradually fill in soil behind the board, gravel in front of the board, pulling the board out every few inches.

**Step Seven** Finish backfilling around the cellar sides. Later you may need to fill in with more soil as the disturbed ground settles.

**Step Eight** Using construction cement, attach rigid two-inch sheets of urethane or Styrofoam to the exterior surface of the root cellar roof.

**Step Nine** Spread a sheet of six-mil polyethylene over the insulation.

**Step Ten** Cover the root cellar roof with two to four feet of earth. Plant grass on the roof to hold the soil so it doesn't run off in the rain.

**Step Eleven** Spread gravel on the root cellar floor (optional).

**Step Twelve** Make shelves. A common mistake is to construct shelves so that they touch the wall. It is much better to leave a two-inch gap between the shelf and the wall, to allow for air circulation. Brace the shelves with scrap wood as shown. We think it is also better not to make shelves too deep. Sixteen inches wide should be plenty.

## Perimeter Drainage

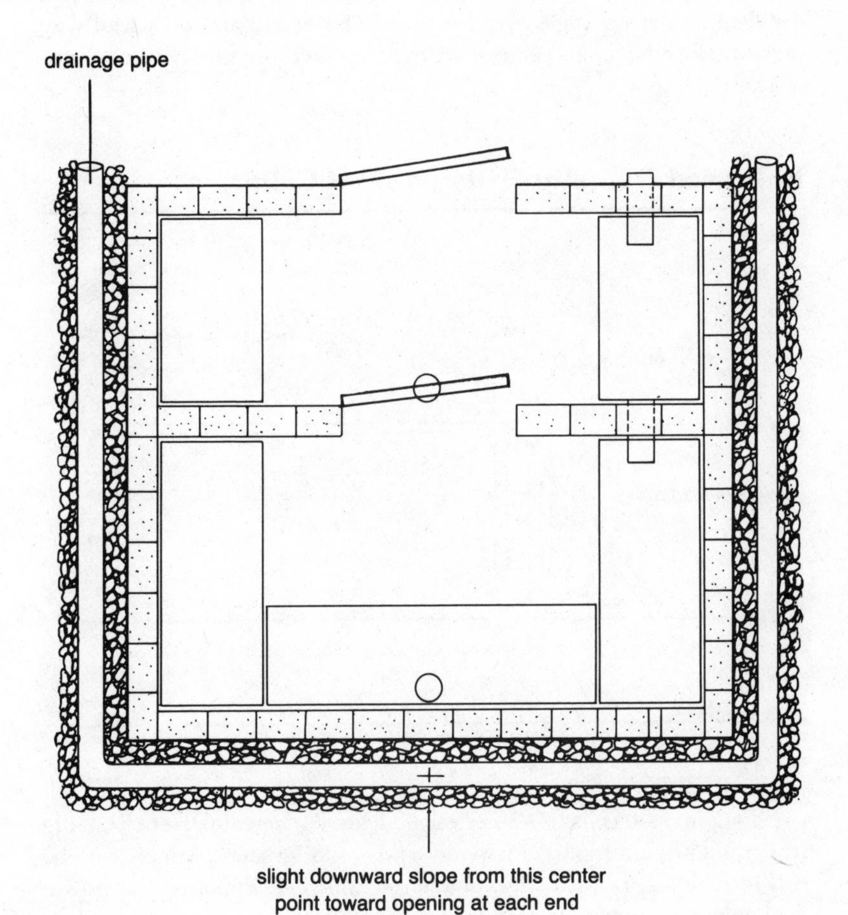

**Step Thirteen (Can also be done earlier.)** Insulate a standard door as described on page 158, and hang the door.

One man we talked to had trouble with the door of his in-hill root cellar freezing shut. (Perhaps the drainage was not very good.) He solved his problem by adding side wings and a cellar door bulkhead, installed at a steep angle over the door. This would also be a neat way to give the cellar more protection from weather extremes without adding an anteroom.

## Bulkhead Doors to Hillside Root Cellar

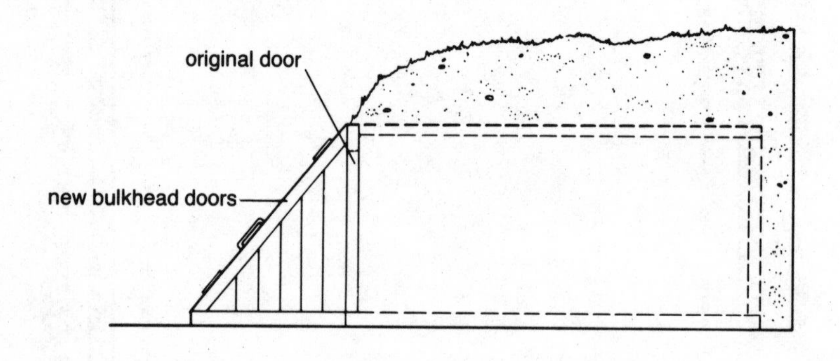


Dug-in root cellar #2 is designed for flat terrain. Here you dig straight down under level ground and build in steps, which can be concrete blocks, railroad ties, or poured concrete. Preformed steps are available, too, from concrete block dealers. Note that a drainage trench (to the right of the air vent pipe) should be dug along the back of the root cellar, at ground level, to drain off rain water. Direct the trench away from the cellar. The French drain is simply a pit lined with gravel.

## Dug-In Root Cellar Number 2

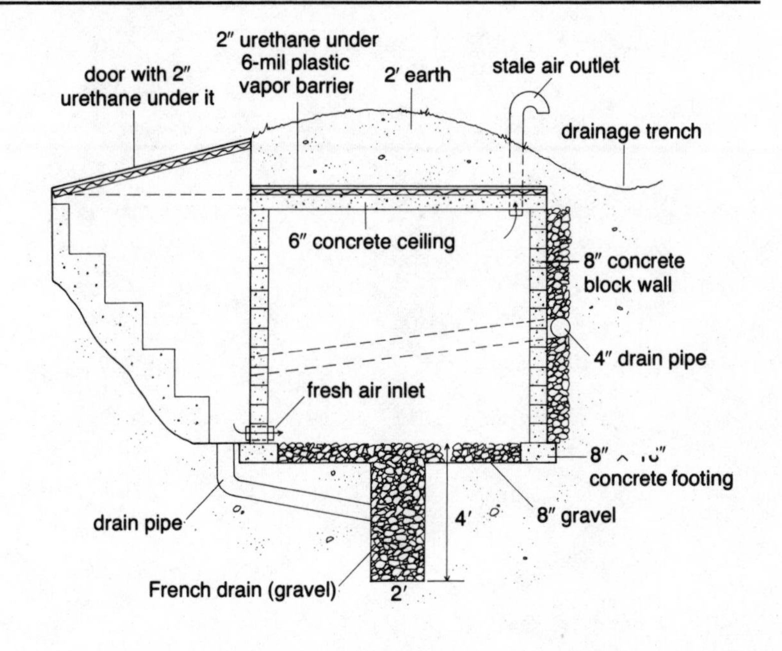

# SECTION FIVE

## "Here's What We Did. . . !"

# 17

# Root Cellaring Experiences

There's something very comforting about being surrounded by a winter's food — stepping around full bins in the cellar and ducking under swaying bunches in the attic, touching aromatic herbs whose fragrance at times fills the house. We all benefit. The whole family works together to bring in the harvest, which we enjoy at leisure during the cold months.

Patricia Earnest
"Storing Food for the Winter"
*Organic Gardening and Farming*

Cabin Creek, Montana, can be a cold place in winter. During the 33 years the Heide family lived on Cabin Creek Ranch, the temperature plummeted to 60 degrees below zero F twice, but their well-built underground root cellar remained frost-proof. The Heide's cellar, which they built in 1901, was a 12-by-14-foot cell dug 7 feet underground. They 'dozed out the hole in the old-time way — with a horse and scraper. The sides of the chamber were framed with cedar poles and scrap planks and the whole building was covered by a grassy knoll. A shingle-roofed entryway supported the outer door and then, several steps down, there was a heavier inner door at floor level. The roof vent was a 1-foot-square wooden chute extending 3 feet above the dirt roof. In cold months this was stuffed top and bottom with old gunny sacks. "Frost-free, fool-proof storage," Mrs. Heide exclaims, "*unless* you forget to close the door!

## Heides' Root Cellar

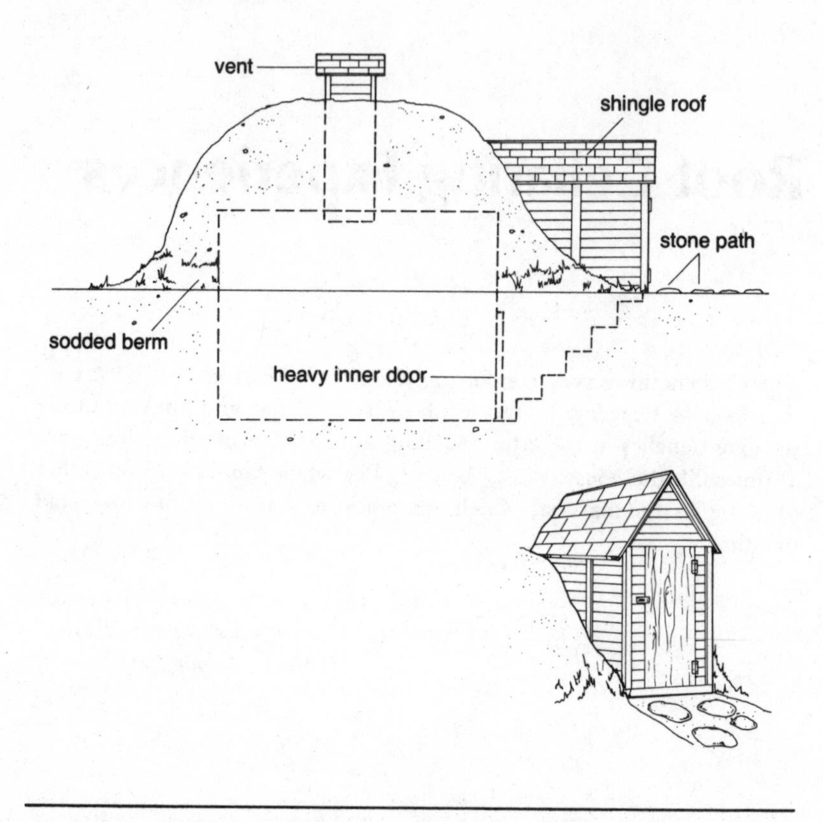


"We were 15 miles from the nearest store," Ella Heide remembers. "We kept vegetables, dairy products, even hams, bacon, and cheese in the cellar. In fact, there was nothing that wasn't stored there.

"The root cellar was cool enough in summer for us to store cream for making butter for a fancy clientele in town. From 1917 to 1934 we had a once-a-week delivery route, via buckboard and driving team, to a town where we sold cream, buttermilk, frying chickens, and root vegetables in season, especially fresh-dug April parsnips. *The root cellar was the most crucial link in the subsistence chain for a good unworried life.*"

Our conversation with Mrs. Heide set us to thinking. Home canning has been common practice for something over 100 years, freezing for perhaps 40 years at most. We consider these technologies to be conveniences, and of course they are. Now, we have no wish to turn back the clock. We're very glad to be living here and now. But haven't we been missing out on a truly basic convenience — the practice of root cellaring — in our preoccupation with jars and lids and blanching kettles and freezer bags? It's as though we've forgotten briefly, almost momentarily, considering the long sweep of human history, how to make use of natural rhythms, how to sensibly meet and participate in each season of the year, how to put natural cold storage to work for us. Now we need root cellars again. Perhaps, in a way, more than ever.

Building a root cellar is work, no doubt about that. But once it's done, keeping food there is simplicity itself. Root cellaring, we're convinced, is more than a forgotten art. We'd say, rather, that it's a vital, freeing tool for the good life — now and in the future.

But you needn't take our word for it. On the following pages you'll find real-life accounts of root cellaring techniques that work. We interviewed city, town, country, and suburban folks — most, but not all of them, gardeners — to see what people really were doing to keep food for winter. We wanted to see how plans, site, and materials came together in actual practice. We hoped to find some creative improvisors, and we did. We tried to find out what foods people were keeping, how they had worked out their food cellars, what problems they'd had in building and operation.

Here, then, is a look at the backyard, understairs, underground abundance enjoyed by a varied group of good people. We hope that their ingenuity and vitality will give you some helpful clues for planning your own root cellaring system.

## Root Cellaring in the City

Over a period of years, Harriet and Howell Heaney have turned their city lot into a productive and eye-pleasing homestead. Fruit trees and berry bushes line the edges of their long, narrow backyard. A wide perennial border leads to a patchwork of compact, intensively planted vegetable beds at the end of the yard. A disciplined row of well-tied tomato plants flanks the border fence. Since the garden is kept in production all season, well into the fall, the Heaneys have basketsful of fine, long-keeping vegetables to tide them over the winter.

Their experience is proof that it's not hard to put vegetables by, even in the city, if you know how. These clever Philadelphia gardeners have built a cold box in their basement to store the produce from their backyard garden. This elegantly simple solution to the storage space problem could be adopted by anyone with a basement window. Most of the box can be made of recycled materials, so this is a money-saving, resource-sparing trick as well.

The Heaney's winter keeper is a double-walled box with fiberglass insulation stuffed between the two layers of wood. The box is built to fit closely around the window opening. It rests on the foot-thick wall of the old stone house and extends one foot into the basement room. Bookcases on both sides help to support it. (In a house with thinner foundation walls the box might need wood struts or angle irons underneath to support its weight.)

The large rectangular box holds two refrigerator crispers full of fresh produce. Plastic bags around the vegetables retain humidity. A hooked-on door shuts the small closet off from the warmer basement room.

Operation of the cold keeper is as simple as its construction. In the fall, the window is opened to allow cool night air to chill the compart-

## Heaneys' Cold Box

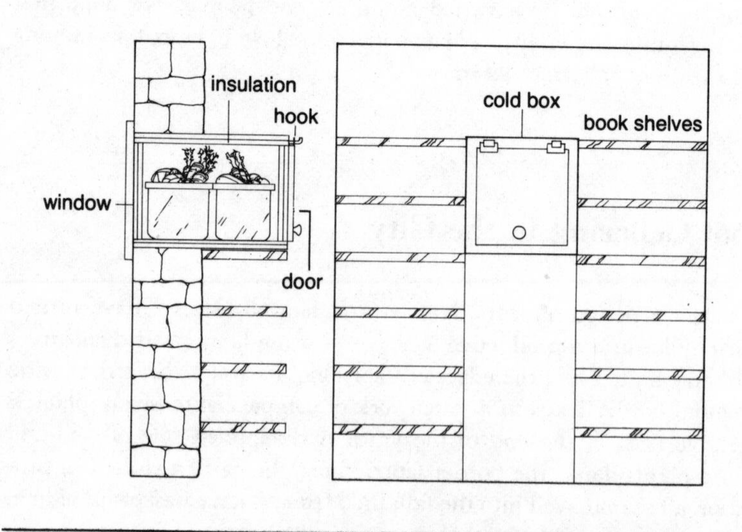

ment. The window is kept closed in winter, when enough cold air enters the box to keep the temperature cold but not freezing. The Heaneys have not recorded temperatures in their cold box but they have successfully stored carrots, cabbage, lettuce, fruit, and cheese in the space. The only problem they've encountered has been rodent damage when the window was opened, but they solved this easily enough by installing a hardware cloth screen over the window opening.

The dimensions of the box can be custom-planned to fit the space *you* have. You could even construct a temporary — but adequate — cold box by nesting a cardboard carton inside another slightly larger carton with one to two inches of insulation filling the gap between cartons. If you prefer not to buy your insulation, you can use sawdust, leaves, or those puffed plastic chips in which vitamin supplements are often packed. If your basement windowsill is narrow, you could suspend the box from the ceiling joists with wood struts or steel strapping tape. Any kind of scrap lumber may be used to construct such a box. The important things to remember are to insulate the box, let in cool air, keep out competition, and enjoy the vegetables! And congratulate yourself for managing to accomplish so much in such a small space, with simple materials.

## Double-Purpose Basement Stairs

Proof that ingenuity and common sense can find good storage spots for vegetables in almost any setting can be found in the basement of Eleanor and Elbert Kohler's snug Cape Cod house. The Kohlers grow many vegetables in the garden which occupies the entire side yard of their small lot. They've lived in this house, on a quiet street in a large town, for over 40 years — time enough to develop a system! In their 80s now, they still enjoy working together in the garden and preparing their produce for winter use. On the early April day when we visited them, Mr. Kohler had part of his garden dug and raked and had already planted a wide row of onion seeds in a specially prepared raised bed. And he still had potatoes and onions in good condition in his basement storage drawers.

The drawer storage system the Kohlers have worked out has proven to be a convenient and easily managed kind of small-scale root cellar. They don't need great quantities of vegetables. There are just the two of them — they are small people and they eat small meals. But they like to be independent and well prepared.

The Kohlers' vegetable storage bins are long drawers built into the steps leading from their basement to the back yard. Mr. Kohler made the drawers out of scraps of one-inch wood. This beautifully simple plan allows them to retrieve vegetables at their convenience regardless of outdoor weather. At the same time, the natural refrigeration of cold air descending into the cellar cools the storage bins enough to keep root vegetables in good condition. On bitter cold nights, Mr. Kohler opens a vent in the inside basement door to permit warm air from the heated basement to keep the temperature of the steps above freezing.

Basement step root storage—the drawers are filled with potatoes.

## Kohlers' Storage Drawers

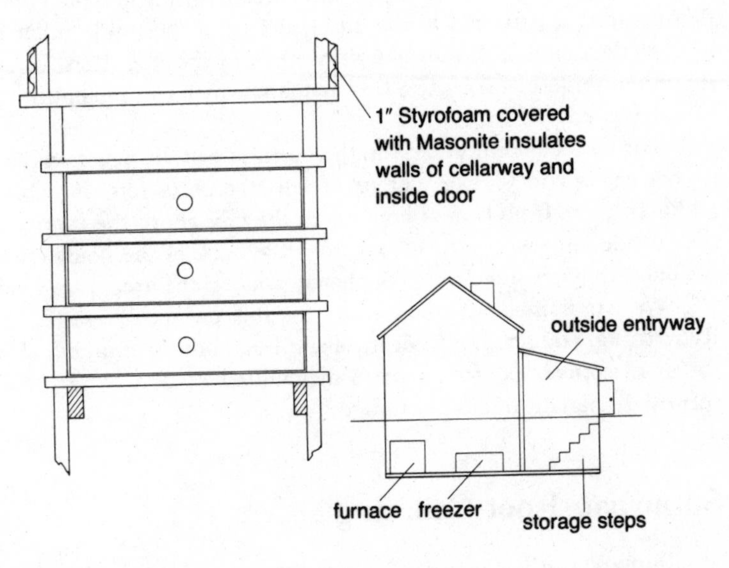


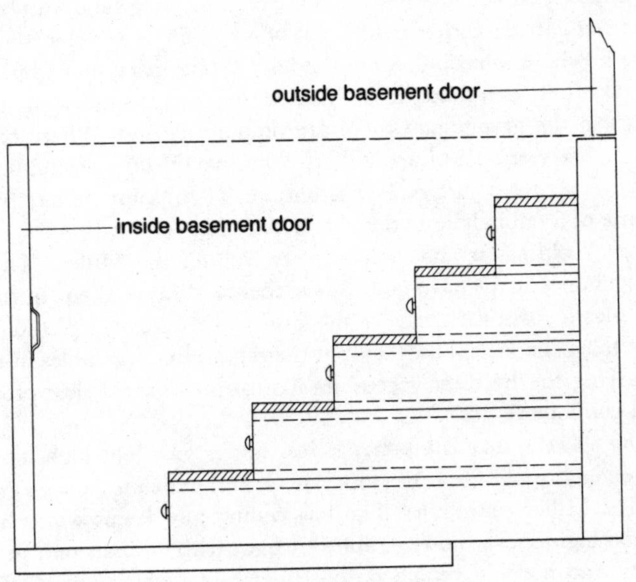

The outside door and the ceiling and walls of the cellar entrance are insulated with one-inch Styrofoam covered with Masonite panels. A thermometer is attached to the outer surface of the inside cellar door.

As the sketch indicates, the drawers form the step risers. They are supported underneath by two two-by-fours, which run the entire length of the drawers.

The basement entrance is on the north side of the house, so the area is cool enough in October for early fall storage of root vegeables. In addition, there is no heated house area directly above the storage bins; the outside entryway extends out from the back of the house. A small shedlike structure attached to the house protects the area from weather.

We can't imagine a better use for this ordinarily wasted space. Retrofitting, you see, need not involve a large outlay of time and money; small corners like this, used to their fullest capacity, can provide a surprising amount of handy storage.

## Suburban Root Cellaring

Suppose you live in a newish house on a nice big suburban lot. You plant a big garden and take off a lot of good fall vegetables, but there may be nowhere to keep them in the house. The space that would have been cellar has been made into a heated, finished family room. There's no room in the garage and there are no outbuildings. Where can you stow your harvest? The Earl Millers, who live outside Mechanicsburg, Pennsylvania, have just such a situation. Their solution has been to make use of a cubbyhole under the back porch.

This "cold cellar" has worked very well for the Millers. They dig root vegetables in October and November and layer them in sand in simple plastic foam ice chests — the kind you'd take on a picnic. When they've filled the ice chests, they put the remaining vegetables in plastic pails, saving the hardiest vegetables like turnips for this less protected type of container.

The space under the porch is just under four feet high, but that allows enough room for a person to kneel, and crawling in is good exercise, Mrs. Miller maintains. The low ceiling also keeps warmth from rising too high above the vegetables. A light with a chain pull provides visibility and it could be left on on super-cold nights to give off some heat. Penny the dog, who at age 16 needs a little extra consideration, has a bed under the porch too and no doubt her body heat helps a bit on cold nights to moderate the temperature. The bare dirt floor keeps the

Mrs. Earl Miller made imaginative use of this under-porch storage area to keep her family in winter vegetables.

space damp, as it should be to prevent shrivelling of the vegetables. Batts of fiberglass insulation have been inserted between all the studs around the three outer walls of the under-porch space. Aluminum clapboards form the exterior siding.

The Millers keep carrots, kohlrabi, and turnips in plastic foam chests in their vegetable cubbyhole. Packed in damp sand, the produce keeps until spring. When we visited in December, the turnips were still as smooth and firm as though they had just been pulled. Carrots, harvested a bit earlier, had put out some sprouts because the fall weather had been unusually mild well into December, but the carrots were still crisp and hard, not at all pithy. The kohlrabi, picked small as it should be, was still crisp.

The Millers have not recorded temperatures in their storage spot, but they did say that potatoes kept there in a heap — unprotected by a foam chest — froze during bitter cold weather, so it appears that some sort of insulation in the container, or at least a covering of old coats or blankets, is necessary in such a set-up to keep the food from freezing.

As we drove away from the Millers' productive lot, past the strawberry bed, the big plowed garden, and the row of fruit trees, we thought of the many houses we'd seen that had similar spaces under porches or front entryways, houses both old and new, in towns and rural areas. All they'd need for a good serviceable cold storage area would be a few batts of insulation, perhaps a light, and possibly a good shovelling out of accumulated treasures. Still more proof, to us, that with determination and imagination you can often find or modify existing space to keep the good things you've grown.

## Where There's a Well, There's a Way

When Francis and Kathleen Olweiler built their own house of native stone, they also made mortared stone walls to line the sides of the pit in which the pressure tank for their submersible well pump was

Apples stored in the Olweilers' well pump house.

## Olweilers' Well Cellar

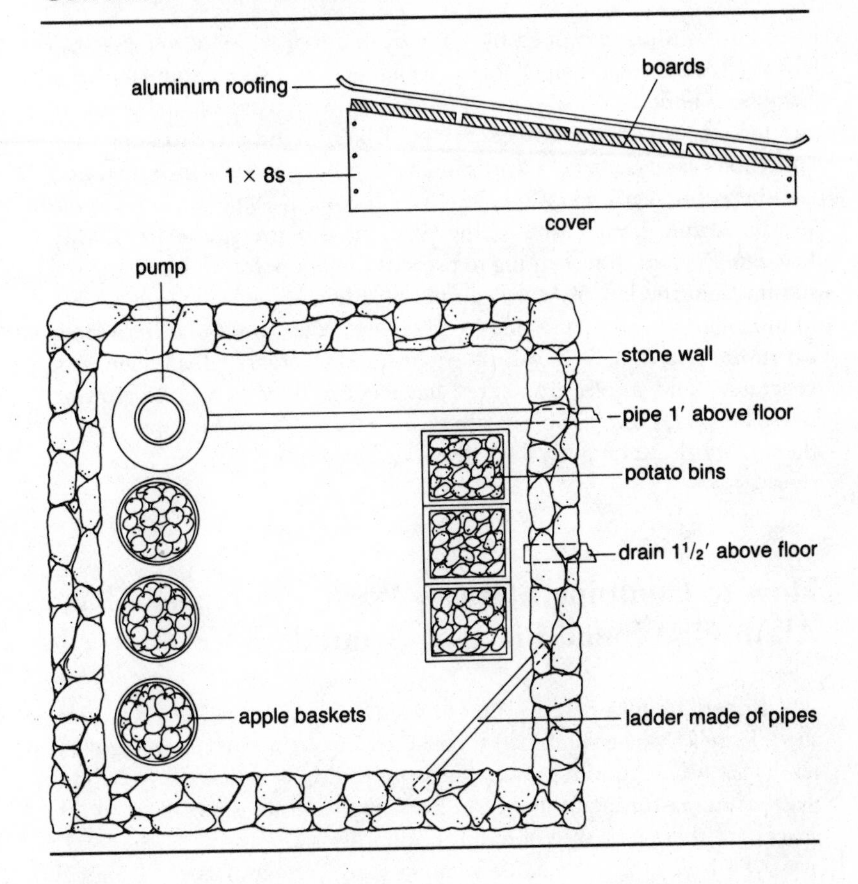


housed. It didn't take them long to discover that they had an excellent root cellar right there. The pit is cool and damp — just right for storing apples and root vegetables. In fact, Francis says, "We'd be taxed to find a better place for us to store vegetables here."

The well pit measures six-by-six feet inside, and it is about five feet deep. The pump pressure tank sits in one corner, with a pipe a foot above ground level leading from the tank to the house. A two-inch drainpipe has been set in the stonework on one side of the wall. A thermometer hangs near the pressure tank. The access ladder Francis built is particularly clever: three pieces of one-inch scrap pipe are mortared into the stonework at an angle across one corner. Neat, simple and easy to use. The floor of the pit is earth, covered with gravel. Even with the well cap propped up, a fine appley aroma pervaded the area.

The well pit is capped by a removable roof made of four two-by-eights (two cut on a slight slant) to which one-by-two-inch furring strips have been nailed. A final layer of aluminum roofing is fastened on top of the furring strips. A sturdy cedar pole is kept handy to prop open the cover for easy access. The fruit and vegetables are kept in bushel and half-bushel baskets, set on the floor and also on the pipe attached to the pressure tank. Trees around the well pit provide shade from May through October, thus helping to prevent the temperature from straying too far from the low 50s even in mild weather.

The Olweilers tend a small vegetable garden, mostly for fresh foods in summer, but they buy large quantities of organically grown root vegetables and apples from area farmers for their family of growing boys. This pit, then, has proven to be a useful cool cupboard, permitting them to stock up on good food three seasons of the year.

## How to Combine Storage, Well, and Root Cellar All in One Small Piece of Ground

Robert Loupin had a well. He needed a root cellar and a storage shed. Here's how he combined them: First he dug a hole about eight by ten by six feet around the well. The water pump is beside the well. The water running through this system helps to moderate the cold air in the space and there is plenty of natural humidity.

"To wall in my vegetable-keeping space," Loupin says, "I hauled flat rocks from potato fields. They stack very easily without the use of cement. All you need is a strong back, a good spade, and lots of rocks. It took me about two weeks to lay the rocks. Then I ran a cement form around the edge of the walled-in space to support a 10-by-12-foot metal storage shed I had. I enter the root cellar from inside the building through a hole cut in the floor."

How does it work? Without measuring the humidity, Loupin knows it is good and damp down there. On a late December day the topside temperature on Loupin's Lincoln, Maine, place was 10 degrees F, but in his root cellar the temperature measured a steady 36 degrees in December and January.

The floor is dirt. The well is not covered, so it gives off some heat. If it were not for the moderating influence of the water, some ceiling insulation would be necessary. All the produce is set on pallets. Carrots packed in sand and potatoes in bins were in fine shape. And apples, Loupin exults, "are just as good as when they were picked!" He keeps the apples and potatoes in separate corners of the cellar. The apples (picked October first) are Cortlands, individually wrapped in tissue or newspaper.

## Loupins' Root Cellar

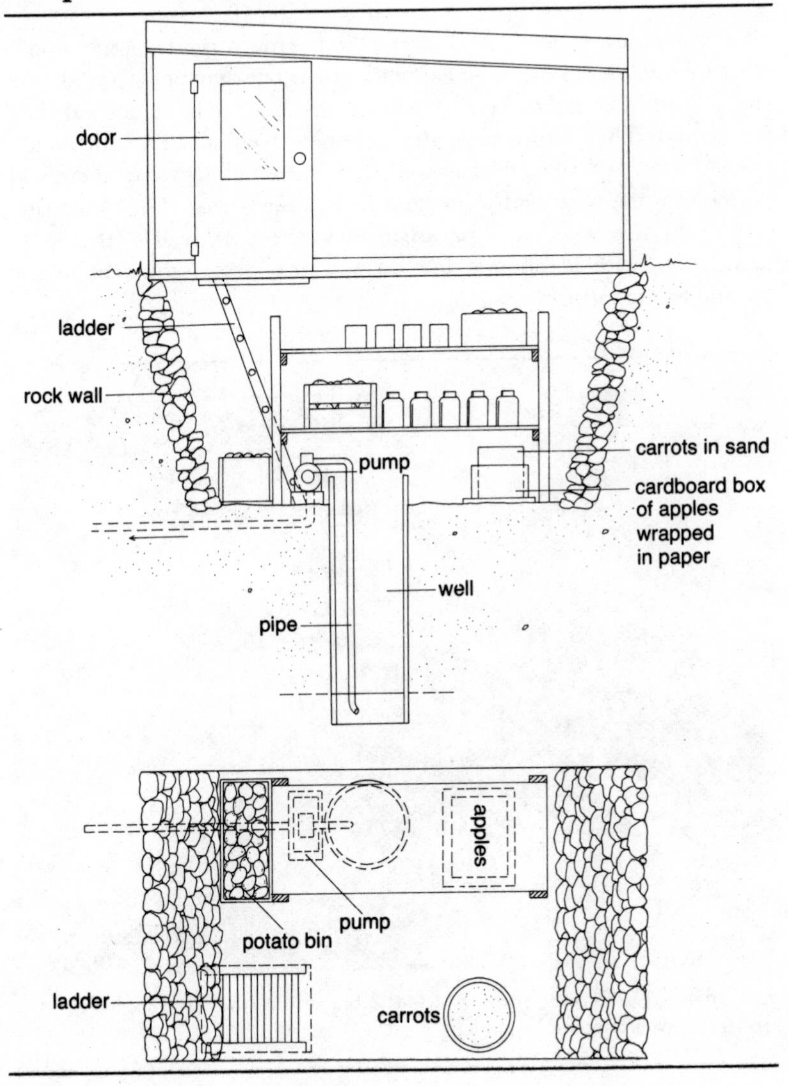

How's that for four-dimensional use of a single 10-by-12-foot plot of ground?

## Putting in an Underground Root Cellar Adjoining an Established House

Harry and Nancy Graver were able to grow a good storage supply of root vegetables in the large garden on their one-acre place outside the city of York, Pennsylvania. But they needed a place to keep what they had raised. They had a cool, dry basement well stocked with home-canned fruits and vegetables as well as wheat and honey, but there was no room in the basement to make a root storage area. Harry's solution was to dig a root cellar room adjacent to the back wall of the house foundation, with a separate outside entrance door right next to the outside basement door.

Entrance to the Graver root cellar—the basement door is at right, steps up to the patio on the left.

It wasn't quite as easy as it sounds, though. In order to get at the soil that was to be dug away, Harry had to use a pick and shovel to break up a concrete patio in that spot. The patio was cracked and in need of repair anyway, but it was still hard work to do away with it.

With the former patio finally out of the way, excavation could begin. Harry had a friend who owned a backhoe and was willing to trade his labor for help Harry had already given him. The friend dug a hole eight feet deep and just over eight by ten feet wide, taking great care not to disturb the house foundation or the already existing cellar steps.

Then, right after the hole was dug and before any further work could be done, a week of heavy rain partly filled the hole with mud. (Luckily, drainage was excellent; slanting vertical shale is under the root cellar.) "Back to square one!" Harry thought, but he got busy with his shovel and hand-dug all the sluiced-in mud out of his cellar hole. Then, with an eye on the sky and an ear out for the weather reports, he went ahead with the pouring of a footing 6 inches deep and about 13 inches wide, mixing the concrete himself and pouring it in forms he'd built of scrap boards. After the footing had cured, Harry and a helper laid up nine courses of concrete blocks to form an 8-by-10-foot room, leaving a doorway. For added strength, they made a poured concrete wall next to the house foundation, using a stone-filled form and pouring concrete over the stones.

Backfilling the space between the root cellar walls and the sides of the excavated area was done by hand — the only practical course for such a small area. Harry tamped the soil firmly every eight inches or so as he filled it in. He is pretty sure that there are still a few air spaces left but they don't worry him; he considers them insulating features. But too much air incorporated in loose fill soil would, of course, be likely to cause cracking and collapse of overlying concrete if the topsoil subsided into the air spaces.

Harry capped the cellar with a layer of poured concrete over 3 inches of Styrofoam insulation he'd picked up on sale. Here's how he did it: He laid sheets of 4-by-8-foot exterior plywood (½ inch thick) over the top course of the concrete blocks, but not covering the holes. The plywood was reinforced from beneath with oak two-by-fours nailed on every 2 feet. To keep the three sheets of 1-inch Styrofoam in place while the cap was being poured, he wired them together. Reinforcing rods ⅝ inch thick were embedded in the poured concrete cap every 16 inches. The structure was further stabilized by pouring concrete into the holes in the concrete block walls while pouring the cap, thus solidifying the wall to resist the pressure of the surrounding soil.

Harry purposely did not tie in the oak ceiling supports to the cement structure. They only served the temporary purpose of supporting the poured concrete.

Another precaution Harry took when pouring the root cellar roof/patio surface was to install metal expansion joints at six-foot intervals in the concrete. Expansion joints direct the cracking that often results from uneven expansion, so that instead of a random, jagged crack in your concrete surface, you have an acceptable, straight-line crack. Felt strips are also used as expansion joints, but they absorb moisture and deteriorate, leaving a larger space for soil to wash in and weed seeds to take root. Felt also tends to expand less evenly than metal, and some workers have told us that it is less effective than metal in holding down the concrete so that it doesn't rise excessively.

Harry was glad that he did put in the expansion joints, because one section of the concrete cap rose a good two inches during the record cold

The Gravers' metal expansion joint.

winter that followed construction of the root cellar. (The poured concrete surface was made larger than the root cellar roof area in order to form another patio to replace the one that had been torn up in order to dig the root cellar underneath.)

The root cellar is easily accessible by way of the cellar steps, or from inside by going out through the adjacent cellar door. The Gravers appreciated this convenience during the heavy snow that blanketed their area the winter after the root cellar was built.

When you enter the cellar you must stoop a bit to get through the door, but that is easier than using a ladder. Two concrete block steps lead you down to the gravel-surfaced eight-by-ten-foot room. We visited the Gravers' root cellar at the end of March — not the best time, of course, to get the full picture of the abundance that fed the eight-member Graver family all winter, but a very good time to see how well the remaining vegetables had come through the winter.

Potatoes were in excellent condition — firm, sound, and still unsprouted. Carrots — and they were big ones — were still sound and hard, with a few small leafy sprouts. Turnips had put on enough pale, leafy tops to make a nice panful of stewing greens, but they were still firm and shapely. Beets — obviously the small ones that had sifted through when the cook chose the larger ones for the evening meal — were somewhat shrivelled but still usable. All vegetables were stored in crates and baskets without any surrounding packing material. Two bacons hanging from hooks in the ceiling added to the feeling of homey preparedness.

Harry had made a quickie shelf by setting pine two-by-twos on concrete blocks to keep the produce up off the very damp floor. (Oak shelving, Mike notes, would last longer in this damp space.) For the second year of root cellar operation, Harry constructed more elaborate shelves to make better use of vertical space. His method of stabilizing the shelves is one of those simple but effective gimmicks that should be shared.

The shelves are made up of two-by-two uprights with one-by-ten shelf boards supported by nailed-on one-by-twos. (See diagram.) In order to prevent the shelves from pitching forward, Harry makes the front uprights longer than the back ones. In this way the center of gravity on the shelves is far enough toward the back to hold them in place.

At the same time, it is important to maintain air circulation between the shelf and the wall to avoid mold build-up. That's why the supporting one-by-two on the top is cut longer — the extension holds the shelf away from the wall.

The Gravers' root cellar maintained a temperature of 40 degrees F all winter long — despite prolonged cold days during which daytime temperatures remained in the 20s and night temperatures dropped to 10 degrees F or even lower. If extreme cold weather would threaten to cause a freeze in the storage area, Harry would drill a ½-inch hole in the concrete block wall and run an electric wire from the house into the root cellar, with the switch in the house, so that he could put a light bulb in the cellar which could be left on in frigid weather.

The Gravers were wise to include Styrofoam insulation in their root

Interior of the Graver root cellar with hanging bacons and turnips, potatoes, carrots, and beets in bins on the gravel floor.

# Gravers' Root Cellar

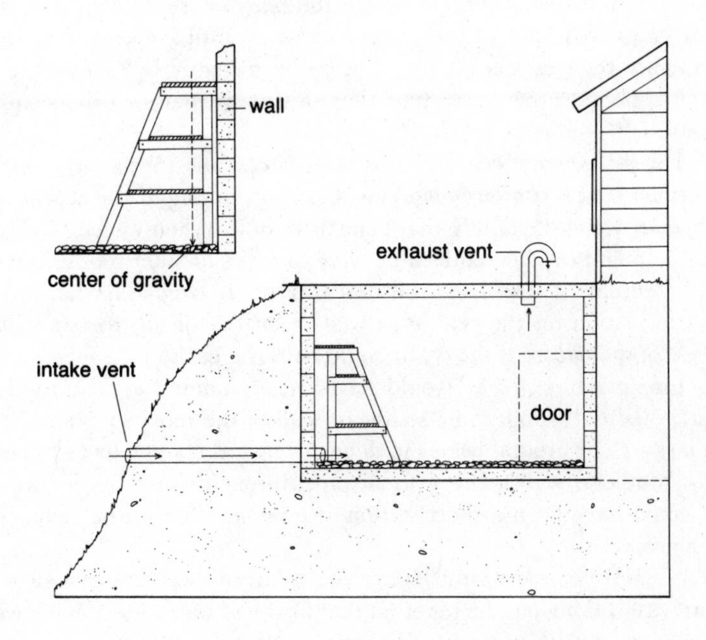


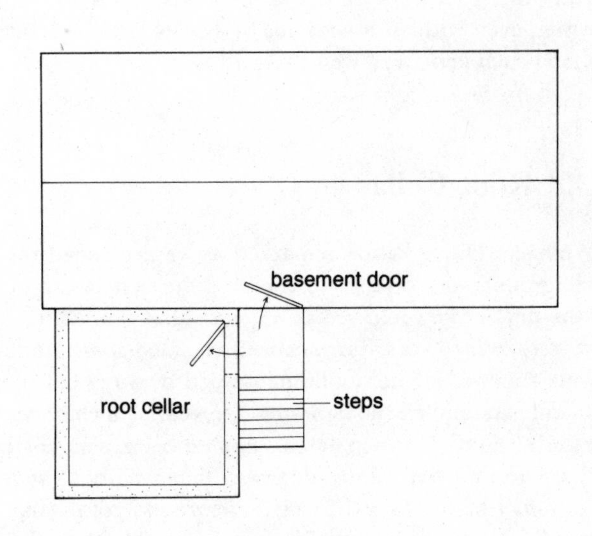

cellar ceiling because the cellar is built on the south side of the house, where fall and spring sun can warm the place. Fortunately, most root cellar crops may safely be left in the ground until November in their area; since their root cellar may not cool sufficiently by October to be used for storage, they may find that this extra margin will be quite helpful.

The Graver root cellar is neat, well made, and ingeniously tied in to the house. It is obviously working and contributing to the health and independence of the family that benefits from it. When we asked Harry what he might do differently if he were to make another root cellar, he said "Ventilate!" And it is true that lack of air circulation has led to mold formation on the ceiling as well as on the turnip roots and the surface of the bacon. If the cellar could be taken back to the footer stage in a time machine, Mike would recommend running a galvanized or plastic pipe out through the southeast wall of the room to open air 15 feet away. (The ground begins to slope downward about 6 feet southeast of the southeast wall.) The exhaust pipe, through which hot air would rise, could have been a short section of pipe installed in the ceiling by the house wall.

When we checked with Harry a year later to see how his new root cellar was working out, he reported that he'd had much less trouble with excess moisture this past year. Perhaps, he theorizes, the soil was water-logged that first year from all the rain that soaked in deep in the hole. At any rate, even without a vent, his bushels of potatoes, turnips, beets, carrots, and such kept very well this past winter.

## An Old Root Cellar

"We like," Diane White remarked as we descended the steps into her underground root cellar, "to combine the best of the old with the best of the new." We could see at a glance that the Whites surely had the best of the old — a solid, rock-wall-enclosed space under the neat little white summer-kitchen building shaded by a big old tree.

When Diane pulled the light-chain, revealing a chamber well filled with crates of home-grown potatoes, apples, beets, and carrots, we decided that she and her family deserved that well-built storage cellar.

The 6-foot-high, 10-by-12-foot underground cold cellar had been constructed especially for vegetable storage by old-timers for whom the vegetables they could put by for winter were a matter of life and health, not convenience. It works as well today as it did 100 years ago. The dirt floor releases soil-borne moisture into the air, which helps to keep the vegetables fresh and crisp. The thick stone walls and surrounding earth maintain a low temperature. There was no thermometer, but on a 40-degree-F December day we'd guess it was about 35 degrees down there. A drain in one corner, covered with hardware cloth, allows excess water to seep out. Rain shed by the concrete pavement above the cellar sometimes washed into the cellar and flooded the floor until the Whites corrected the direction of the drainage flow on the soil surface. There's no air vent, but the loosely filled bulkhead doors allow a certain amount of air to enter the cellar and the ceiling is not airtight.

Several contemporary improvements have made this old root cellar even more serviceable. A concrete walk extending from the entrance steps into the middle of the cellar helped to keep feet clean and dry

The entrance to the Whites' fine old stone-walled root cellar is under the bulkhead on the side of this summer-kitchen building.

## Whites' Old-Time Root Cellar

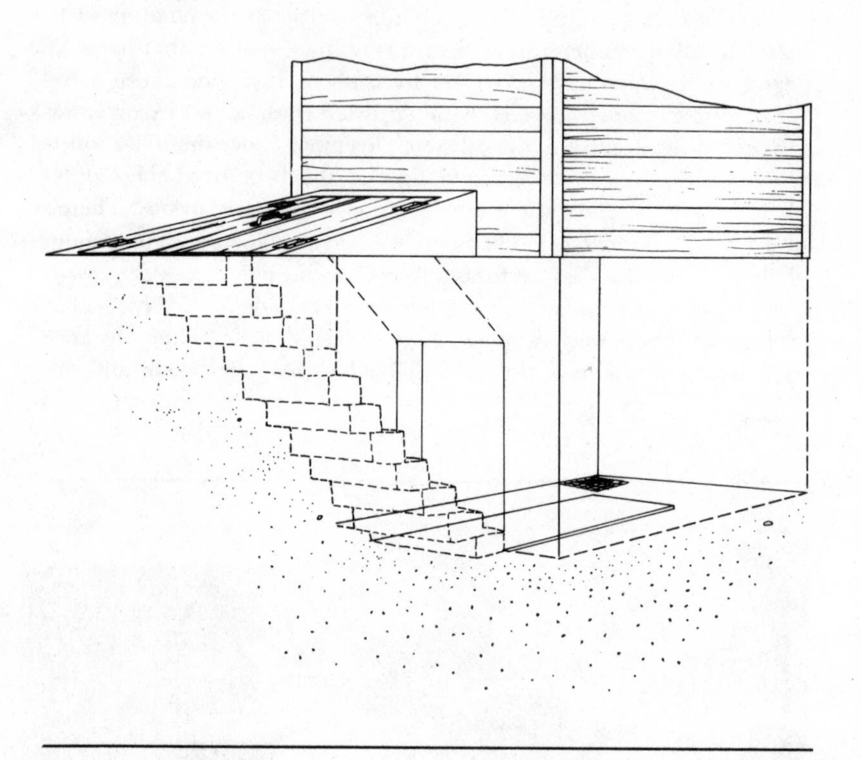

when the earth floor was quite damp, and a layer of fiberglass insulation tacked onto the ceiling blocks the descent of cold air from the unheated summer kitchen above. A light bulb at the doorway illuminates every corner. One more touch would improve the cellar even more. On extra-cold days the cold air descends through the exterior plywood bulkhead doors and threatens to freeze the produce. The Whites plan to install a door at the bottom of the steps to wall out this cold air and stabilize the cellar temperature.

This is mostly a preventive measure, though. They haven't had any problems with their vegetables freezing. Apples and potatoes keep until March. And they keep well together in the same room, Diane notes. Even the red potatoes, which in Diane's experience seldom keep as well as the white ones, remain unsprouted. Perhaps the secret is the constant low temperature. Certainly the presence of apples doesn't seem to have any adverse effect on the potatoes in this cellar.

The Whites have worked out several tricks to keep their produce in good shape. They line all crates with several thicknesses of newspaper and leave a flap of newspapers hanging over the edge. In bitter cold weather, then, it's a simple matter to flip the newspaper blanket back over the vegetables to insulate them. It's also very important, the Whites feel, to keep boxes elevated several inches off the floor so that neither the wood nor the vegetables contact the wet soil. And they have found that apples — especially the early varieties — keep best in large plastic bags which they recycle from other sources.

What a pleasure to see so much good home-raised food in such a fine secure storage space! Although the cellar Diane and Howard White enjoy is a legacy from careful gardeners of another day, the idea could be adapted by anyone who needed a shed, garage, or pony barn. Simply build the small new structure over a concrete block basement. Run

Inside the White root cellar.

cement steps down to the basement, insulate the ceiling, put in a drain and a vent, and raise the shed over joists set across the basement. Dimensions of the cellar (and the building) could be reduced, too. An eight-by-eight-foot cellar would still provide loads of storage room and that might be just the right size for a garden toolshed to store your mower, tiller, rakes, hose, and bikes. Perhaps you could even sneak in a potting-bench corner on the ground-level floor.

## A Planned-from-the-Start Root Cellar in a New House

The owner-designed and owner-built home is becoming more common as people decide to take into their own hands the arduous but satisfying task of building a dwelling. When you design your own house, you can include self-sufficiency features that would not be found in a conventional plan.

The Ephraim Mohrs of Eastern Ontario, Canada, did just that. "After years of false starts, we got a cold room that works," Ephraim reports. "My 1977 carrots were crisp in August 1978 and last year's onions were good until July with a little culling."

The fact that Mrs. Mohr is an architect, "trained to find economical solutions to design problems," as her husband says, helped when it came to integrating the cold room into their custom-built house.

"The basic scheme here," Mohr continues, "is of the under-the-front-step class. This is a general scheme used in a lot of mass-produced subdivisions in these parts, but since the standard front step is maybe eight feet by four on the outside, with inside dimensions just a little over six by three feet, they're really too small to be much use. Besides, this design is often too far out of the ground to get the temperature stability of an underground root cellar.

"In our case the front entry is just one step up from the front lawn. It's a 10-by-8-foot concrete platform that gives access to both the front and back doors. When you go through the front door you are faced with stairs, a half-flight up and a half-flight down. A further half-flight down from the lower level of the house puts you in the cold room under the front entry. The cold room is thus a good size, 6 by 14 feet, all concrete walls and set deep in the ground. Indeed, the air vent has to be in a well in the garden beside the front patio.

# Mohrs' New Root Cellar

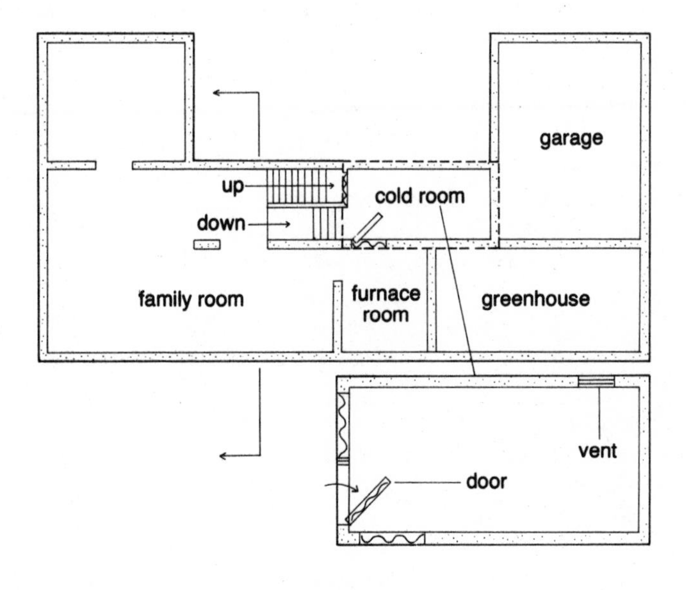


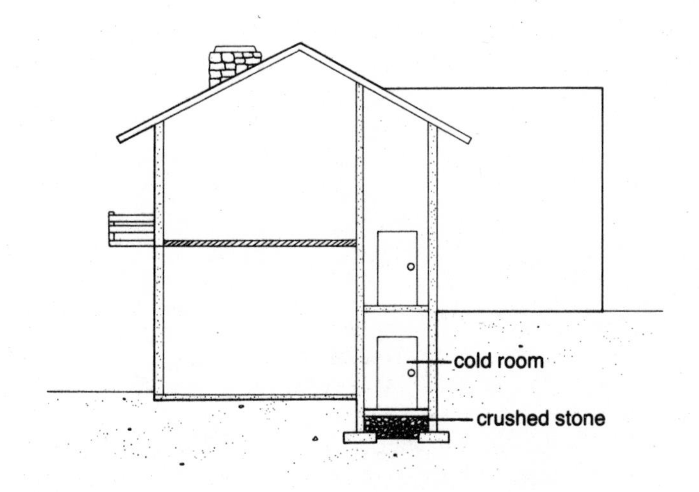

"An unusual feature of our room is that it contains the pump and pressure tank for the well. This provides an extra year-round stabilizer, being at a steady 40 degrees F summer and winter — that's the temperature of the water from the deep well. In summer the water gets a little cooler when we run the hose a lot.

"The first year after we moved in we found ice forming on the concrete ceiling — a great demonstration of how little insulation you get in poured concrete, but a little too cold for me. Besides, that meant that the cold room would get warmth through the ceiling in summer. So, next year we stuck one-inch Styrofoam panels to the ceiling, using construction cement, and also to the top four feet of the exterior walls in the cold room, to make sure that the temperature of the storage area would be set by the earth at its lowest level — warmer in winter, cooler in summer.

"The floor is crushed stone so that condensation on the water tank and any other excess moisture just disappear. Next, the scheme of having the cold room a half-level below the heated lower level means that the cool air does not flow out when the cold room door is opened. (The cold room door is insulated too — more foam plastic glued on.)

"So how does it work? Well, it's great! We use the cold room as a wine cellar, too, although 'round about the end of January it gets too cold in there for the wine and we have to move it out. Beetroot crops love it just over freezing so they keep well. I hang the onions in net bags off the ground. Carrots and dahlia bulbs are packed in boxes of dry peat moss. I keep squash elsewhere. Potatoes kept in sacks on the floor last until March, when the supply runs out.

"It's my experience that the best way to store vegetables depends a lot on the temperature and humidity of the storeroom and the two cannot be separated."

## Our Planned-in Root Cellar

There is plenty of hilly land on our homestead, but it all faces south. Those slopes are perfect for solar greenhouses but we wanted our root cellar less exposed to the sun. Last year we had a good chance to plan a vegetable storage room from scratch when we decided to build a small new house here on our property. Mike designed the house and we built it ourselves, so we were able to do some crazy things like putting a small basement under one corner of the concrete slab-based house. We'll

## Bubels' Root Cellar

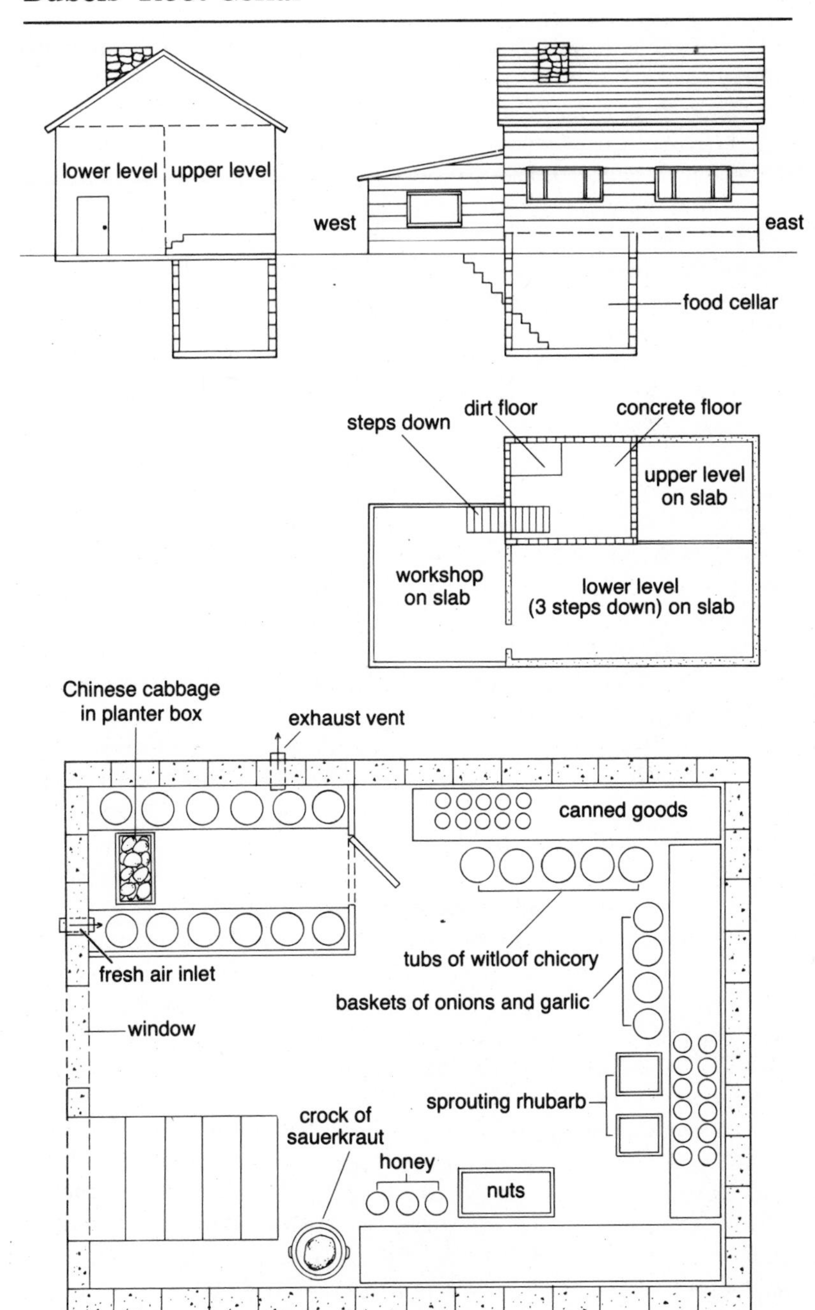

show you in a minute how it was possible to do this and keep the structure sound.

First, though, let's follow the new root cellar from the ground up. The 11-by-13-foot basement area was excavated when trenches for the house foundation footings were dug with a backhoe. (Just to make things interesting, we encountered a huge boulder right at the basement entrance and had to have it blasted so that digging could continue.) Then, after footings were poured, we enclosed the basement area with concrete blocks, leaving a doorway leading to the workshop adjoining the house. The workshop is five steps above the basement and the house is several steps above the workshop.

As you'll see in the diagram, we divided the basement into two zones. The smaller portion is the root cellar. This four-by-five-foot food closet, partitioned off from the rest of the cellar on the northwest corner, has a dirt floor covered with gravel. This little room is damp and cold — just right for storing root vegetables. With shelves on the walls it will hold 15 bushels of food, plenty for the two of us, now that our children are in college. (We were amused to find that our dog discovered too that the root cellar is a good place to store provisions. Molly is a mutt with an appetite for corn. When she can get hold of a dropped ear in the field, she brings it home to gnaw on the front step. Her latest refinement, though, has been to stash several ears of corn in the cellar for future attention. She's considered the cellar her territory every since the summer we started work on the house, when she'd go there for refuge from the heat.)

In order to provide the chimney effect (incoming cool air from a low inlet pushing warm air out a higher outlet vent), we left two spaces in the concrete block wall for intake and vent pipes.

The larger part of the basement is finished with a concrete floor and a drain. This area is not as damp, or quite as cold, as the enclosed root cellar and we expect to use it to store canned goods and some vegetables like onions and garlic that keep well in a cool, dry place. If we had a furnace in the basement, we'd have to insulate our vegetable-keeping room, but since we plan to heat the house with a wood burning box stove in the living room, the basement should remain cool.

We insulated the upper few feet of the basement wall on the outside with the same two-inch sheets of urethane foam that we put around the perimeter of the rest of the house. This helps to prevent severe cold at soil level from penetrating the storage area. Thus the earth's more moderate and constant temperature has more influence on the root cellar than fluctuating surface temperatures.

The ceiling of the cellar, as shown, is made of wood — two layers of exterior fir plywood with a sheet of six-mil plastic between them, supported by two-by-ten joists. The wood ceiling abuts the poured concrete slab that supports the rest of the house. The same flooring will be laid over both the slab and the wood basement ceiling.

In the smaller root storage room we'll put Styrofoam insulation on the ceiling to keep out any warmth from the wood-heated upstairs. Since the room above the root cellar is often kept closed (and therefore unheated) during the peak heating days, we don't anticipate any trouble there.

As you can see, the area is a small one, purposely so. It will be devoted solely to unprocessed vegetables, shelving will be adequate, and the room will be fully utilized. No wasted space, no wasted vegetables. And best of all, no waste of gas or electricity, resources, time, or vitamins in processing. We consider our root cellar a vegetable independence bin, a vital part of our plan for living free.

## Creative Recycling Builds a Root Cellar

Some of the most inspired vegetable storage arrangements are improvised from found materials. It's not always possible, of course, to duplicate another person's lucky find, but we'd like to tell you about one of these ingenious rigs because it shows how alertness to the possibilities of sturdy scrap material can provide a really fine storage bin. Look around — perhaps you'll find something just as good!

Jack Shaull of Cheyenne, Wyoming, devised this handy "winter wannigan box," as he calls it. (The wannigan was the provision shack in a logging camp.)

As Jack tells it, "I happened upon its husky hulk as it was about to be pitched out of a railroad kitchen car. The old slab-sided ice box was being replaced with a new-fangled contraption that plugged into the wall and kept its own cool. The outer shell of my find was made of galvanized sheet metal, about three feet square and four feet deep (no doubt the plans were adapted from the *Monitor* or *Merrimac*). The inside was a sturdy wooden box with a trapdoor lid. A four-inch layer of sawdust was packed between the inner wood box and the metal sheathing.

"I finally got enough earth excavated to bury the massive box, ending up with the trapdoor at ground level. My wife asked me if I intended to trap the neighbor's milk cow. I put up with plenty of teasing about my buried box, but my efforts were rewarded at harvesttime. I had enough room for all the carrots, parsnips, red beets, turnips, and whatever and maybe even the neighbor's milk cow, too, to hibernate for the winter. After filling the box with my garden produce, I threw a sheet of old Celotex sheathing over the top.

"One bright December morning I shovelled the snow off my buried box and opened the trapdoor. All my good homegrown foodstuffs looked as fresh as they did on that October day when I had stowed them away. I felt pretty proud of my salvaged root cellar. In my eagerness to hold up a prize bunch of parsnips (I had noticed my wife watching me from behind the kitchen curtain) I slipped from my kneeling position and plunged head-first into that carrot-filled canyon!

"Retrieving the vegetables is easier now, since I purchased a plastic garbage container. I reach down with a pair of hay hooks and hoist the whole shebang topside, where I can rummage through the goodies like a Yellowstone bear raiding a trash can. And my wife admits it's pretty nice to have fresh carrots and parsnips in the middle of winter."

## It's Never Too Late to Add a Root Cellar

"Since my retirement seven years ago," says A. W. Landry of Lake Lure, North Carolina, "I've had a ball doing some useful things. After about two years of part-time effort — by the 'armstrong' method — I have built a root cellar, and am well pleased with the result.

"Our house is small and has no cellar. We had storage problems, and something had to be done. About 35 feet in back of the house, there is a bank. As it was serving no useful purpose the way it was, we reasoned that it would be ideal for a root cellar. So I started digging.

"Digging straight ahead for about 14 feet, we wound up with an excavated area about 13 feet long and 13 feet wide. I set up small forms and poured each wall — a small section at a time — until there was a concrete wall 8 inches thick and 6 feet, 4 inches high. I poured three walls in this manner, gradually moving the forms higher. The front wall, the fourth, was poured last, leaving sufficient space in the middle to frame in a door. When pouring each side wall, I provided two 3½-inch holes about 4 feet up the wall and about 6½ feet apart, for vent pipes.

"The next step was to backfill the space between the walls and the bank to a depth of about four feet. On top of this dirt and along the

## Landrys' Root Cellar

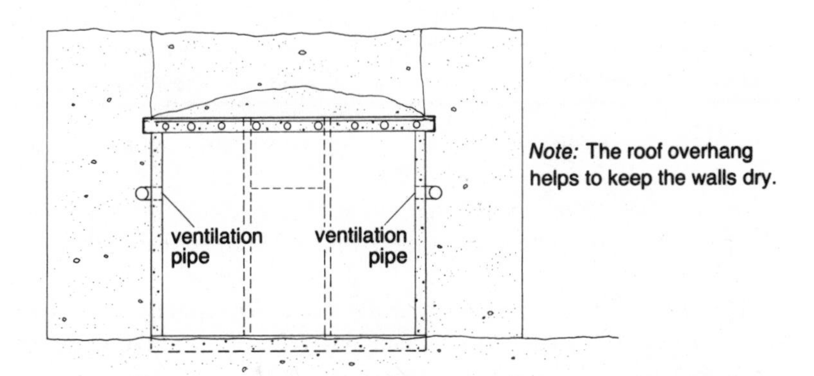


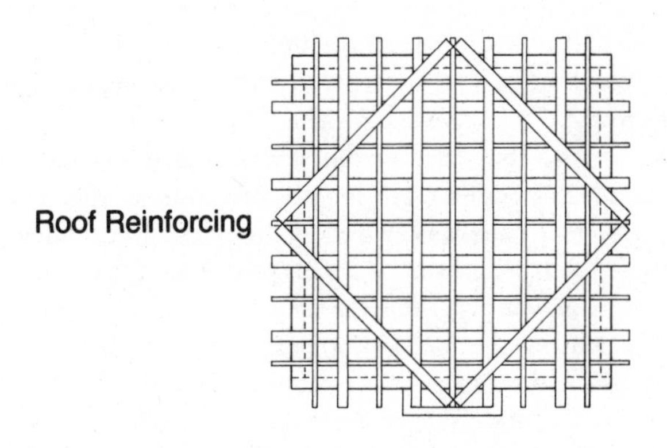


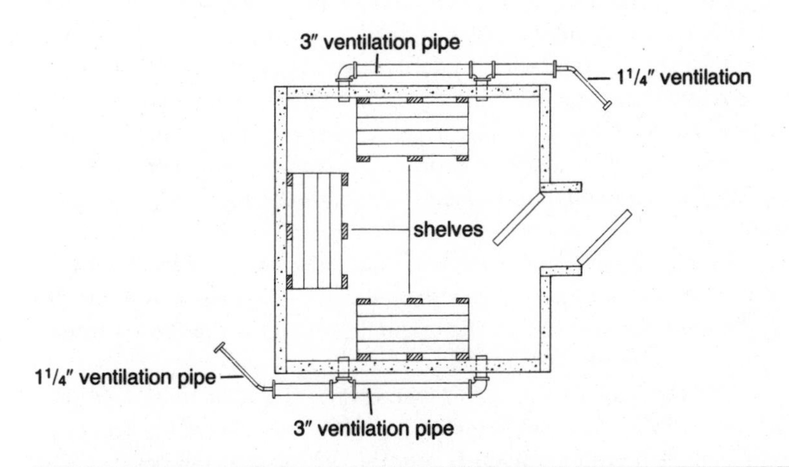

outside of the two side walls six inches from the wall, I placed a six-foot section of three-inch plastic pipe with an elbow on one end and a 'T' connection at the other end. This would be for the ventilation system. I then connected short pieces of three-inch pipe to the elbow and 'T' fitting to go through the walls in the holes provided.

"The last 4 feet of the front pipe and the last 4 feet of the rear pipe on the outside end were reduced from 3 inches to 1¼ inches and provided with an end cap which has an enclosed fine wire mesh screen to keep out flies, bugs, and so on. I then continued backfilling until the soil was at roof level.

"The roof presented a problem for a one-man operation. It had to be strong enough to support a tremendous weight. When completed, the root cellar would be completely covered with dirt except for the doorway. On the roof at the entrance the dirt would be 3½ feet deep, and at the back end it would be approximately 6 feet deep.

"Across the top of the walls I placed steel reinforcing rods — five going crosswise and five lengthwise. These were wired together where they crossed so that they would stay in place when the concrete was poured. Then I added eight lengths of two-inch iron pipe in the same manner — four each way, wired at the points where they intersected. I also placed four lengths of two-inch pipe diagonally across each corner, wired to whatever pipes they crossed.

"In pouring the roof I propped a piece of ¾-inch plywood three-by-four feet and wedged it with lengths of two-by-fours so that it stayed in place on the underside of the reinforcing rods and pipes. At three or four places I drilled small holes in the plywood. Then I threaded baling wire through these holes and twisted it around those pipes on top. Between the props and the wire this form stayed in place while that section of the roof was poured (six to seven inches thick). When one section of the roof had set I cut the wires, knocked the props away, and removed the plywood form. I then set the form up in an adjoining section of roof, poured that, and continued in this way until the roof was completely poured.

"I coated the top side of the roof with a thick layer of tar. On top of this I placed four-mil plastic sheeting, doubled. Then more shovelling to put the soil back and restore the original contours of the land. I intend to plant vines on the sloping soil.

"The inside of the root cellar measures 10 feet wide by 10 feet long. The ceiling height is 6 feet, 4 inches. I set 16 two-by-fours, 7½ feet long, in an upright position, cemented into the dirt floor at the bottom. The top end was wedged against the ceiling. The uprights serve two purposes: as additional support for the roof and to hold the shelves I built.

"I haven't gotten the second door installed yet and the one door is not tightly framed. Still, with the temperature hovering in the teens and low twenties for about a week (unusual for down here), the thermometer in the root cellar stayed around 40 degrees F. Last fall, when we had 65-degree and 75-degree days, the thermometer registered around 46 degrees. Not bad."

## This Family Has Three Root Cellars

Tim and Grace Lefever of Spring Grove, Pennsylvania, are self-sufficiency experts, having homesteaded for many years in a solar-heated house Tim designed long before most of the rest of us caught on to that sensible option. The Lefevers produce their own vegetables organically, growing many different kinds and, like most gardening enthusiasts, trying new varieties each year. They eat well out of the garden all summer and the root cellar takes over in the fall.

According to Tim and Grace, one of the nicest things about having a root cellar is that it provides a cool place to store produce during the marginal days of early and mid fall and again in spring, when warm sun quickly zooms the temperature in even unheated house rooms higher than it should be for good vegetable keeping. When you'd planned to leave potatoes in the ground till late October but heavy September rains force you to dig the spuds so they won't resprout, where do you put them to keep them cool? The root cellar, buffered by the cool surrounding ground, is the best place of all.

"We've been making a lot of sprouts for fresh winter vegetables," Grace said, "but we still depend on our root cellar. In fact, we've installed a fan in one of our cellars to bring cool night air into the cellar to chill it even more rapidly in the early fall."

These vegetable-wise people keep a variety of vegetables in their root cellars — escarole, endive, rutabagas, Chinese cabbage, beets, and more. When we visited them in March, Grace brought up a basket of Chinese cabbage and beets — proof that their system does indeed work.

And yes, we *did* say root cellar*s*. The Lefevers have sold organic produce for years, so they've developed not one but three root cellars. One of the older root cellars on their homestead is in an excavation under an old shed. The vegetable storage area is right above the site of the submersible pump. The water pressure tank sits on the dirt floor. Water flowing through the tank to supply the house helps to keep this cellar slightly warmer in winter. This is especially true if there is a vent to encourage circulation of air.

Another cellar is in the basement under the wall between their house and their whole-foods store. The newest cellar on the Lefever homestead is a concrete block room 6 feet deep, under a metal pole-shed that is used to store organic gardening supplies, which they also sell. The dirt-floored room measures 11 by 12 feet. Above the nine courses of block, there is a poured concrete ceiling 4 inches thick. The ceiling is reinforced with ⅝-inch steel reinforcing rods arranged in a 2-foot-square mesh, wired together where they cross. The area is lit by a droplight. Vegetables are stored in baskets on the ground. The cellar is not vented. Access is by way of a trapdoor and ladder. "To prevent freezing," Grace notes, "we had to leave the hatch door open on the coldest winter days in order to increase the air circulation." An extra piece of plywood and a length of discarded carpet spread over the trapdoor also helped to insulate the area.

Busy not only with their homestead and business chores, but also with a score of environmental concerns and educational programs, the Lefevers somehow manage to do it all, graciously. Surely the home-grown vegetables they eat and keep deserve some credit for helping to make possible all of this worthwhile activity.

## An Alternative to Jogging: Dig a Root Cellar to Keep Fit!

"We had been talking about making a root cellar for several years," said Jean Harper of Greenfield, Indiana, "but due to my husband's full-time job and part-time farming, that was all we had been doing — talking! About two years ago, my father retired. I don't know whether he was tired of hearing us talk, or just wanted something to do. But one day he said, 'Decide where you want this root cellar, and I'll dig it for you.' After we were sure he was serious, we picked a spot close to the house on top of a hill.

"My Dad had just retired from a machine job. Since he wasn't accustomed to hard physical labor, he intended to work on the root cellar three days a week. His plan was to work for three or four hours a

Digging out the Harper root cellar.

day, Monday, Wednesday and Friday, with a break for lunch. He started in June. We didn't give him any idea about what we wanted, because we didn't know ourselves. The first Monday, he did a lot of measuring and staking off, to get the general dimensions. Next came digging up the sod, which I got to use on the bare spots on the front lawn.

"The next Wednesday, right on schedule, Dad arrived with shovel in hand. And so it went, slowly and steadily. By the time he got to the really hard digging, he was in better physical shape. And there *was* a lot of hard digging. Our soil is clay and there are many rocks and tree roots

The Harper root cellar during construction.

in our yard. And it gets very hot down in a six-foot hole in summer! But with a lot of hard work and time spent, the hole grew. Dad worked on the stairway as he dug the main hole, so when he was done with the main digging the stairway was roughly done, too.

"Because we were familiar with concrete block work, and we wanted a permanent addition to our farm, we decided to use blocks for the root cellar walls. So with our much-used concrete mixer, we started pouring the footings. We figured, since we were well below the frost line, that the footer didn't need to be very big. Since fall was coming too fast for us, we got an estimate from a contractor for finishing the job. When we heard that it would cost us $700 to have the contractor lay the blocks and do the stairs, we decided to take the time and do the job ourselves.

"My husband, Jerry, worked weekends, and in a couple of weeks he had the blocks laid. A few weeks later, we poured the stairs and then laid four-inch block leading down the stairway to the inner door of the root cellar. Fall rains and the approaching winter then called a halt to

## Harpers' Root Cellar

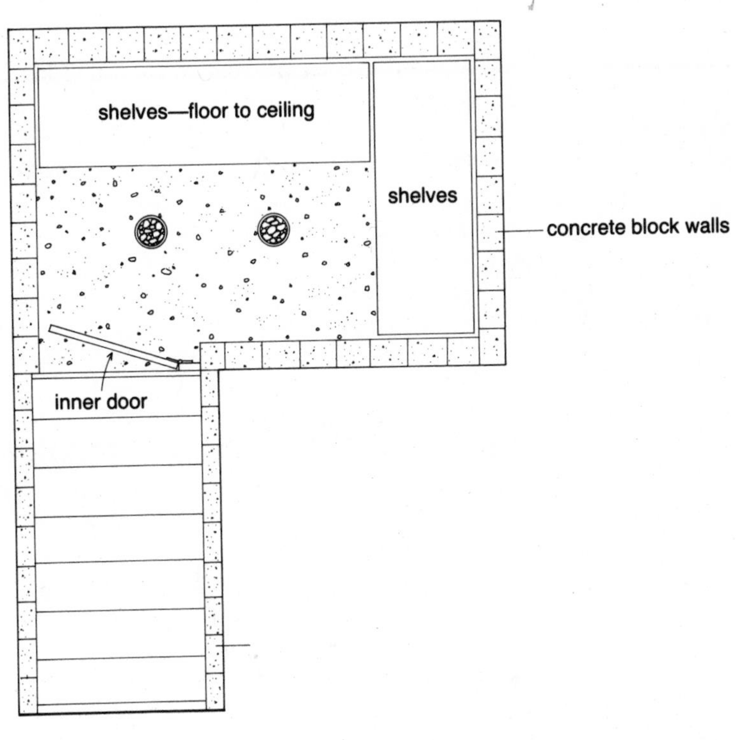


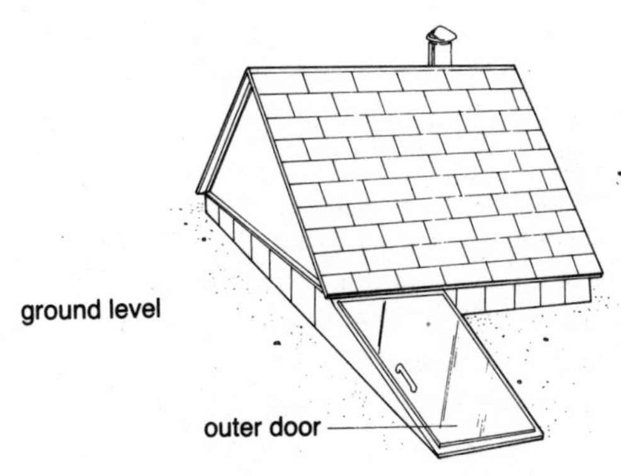

our work. We spread a large canvas over the hole to keep out rain and snow. That was the way it stayed for the coldest winter in Indiana history. Even without a roof, we kept seven bushels of potatoes in the root cellar all winter and didn't lose one potato.

"The next spring, Jerry started on the roof. Using wood left over from other projects and roofing left over from our house, Jerry framed the root cellar roof. He put a vent in the roof. The stairway leading down to the root cellar has a framed metal door, and there is another door at the bottom of the stairs. The floor is dirt, covered with some gravel. We dug another smaller hole in the floor and put in a five-gallon bucket covered with a screen. We figured that we could dip off from the bucket any water that seeped into the cellar. It didn't work out that way, though. Water started seeping in faster than I could dip it out. We solved that problem by digging a trench and laying drain tile for a proper drain.

"We have a small house and limited storage, so we designed shelving to accommodate both bushel baskets and canning supplies. Our shelves are on the north and east sides of the root cellar, from floor to ceiling. We picked up wooden grape crates at the grocery store to hold some of our vegetables. And if we have a question about odors, we store vegetables in covered metal cans. We are planning to get old milk cans to store our apples. We have stored Jerusalem artichokes, potatoes, cabbage, turnips, and flower bulbs. We kept tomatoes until after Thanksgiving and ripened pears. Since this is tornado country, we also keep an old lantern, matches, and a flashlight in the cellar for use in case of a bad storm.

"Although we have not kept accurate records, we estimate that the six-foot, eight-inch-deep, six-by-nine-foot root cellar cost us $200. And we feel that we have made a good investment toward being more self-supporting.

"Incidentally, my father came to me later and said, 'Thanks for letting me dig your root cellar. It added ten years to my life!' The project gave him something to plan for his first year of retirement. And, even more important, the digging helped him to become physically fit and gave him the habit of regular exercise. He is now enjoying an active retirement and we still have more projects, whenever he wants them!"

## A New Cellar on an Old Farm

The Senfts of central Pennsylvania live on a farm that has been in their family for several generations. Their root cellar is a recent addi-

The Senft root cellar is dug into a hill and has an awning to protect the doorway from sun and rain. Note that the stone facings on the sides of the hill surround the root cellar.

tion, though — just three years old when we visited them. It is built to last. When we stopped to admire the structure in early September, several lugs of potatoes were already cooling in the damp dark recess, with early apples soon to follow.

The root cellar is built into the slope of a hill with the door facing north. A 3-to-4-foot capping of soil is packed over the roof of the cellar. An anteroom serves as an air lock to prevent undue warming of the storage area when loading and unloading in mild weather, and undue cooling in severe sub-zero temperatures. The walls of the underground structure are made of concrete blocks which are set on a poured concrete footing. The cellar is six blocks wide, ten blocks high, and seven blocks long, not counting the 3-by-10-foot anteroom. Gravel is spread on the floor. The poured concrete ceiling is arched, with a 4-inch rise from the low points on the side to the highest point in the center. Sturdy, long-lasting shelves made from telephone pole crossarms extend around three sides of the cellar. There is a floor drain in the storage area and two vent

## Senfts' Root Cellar

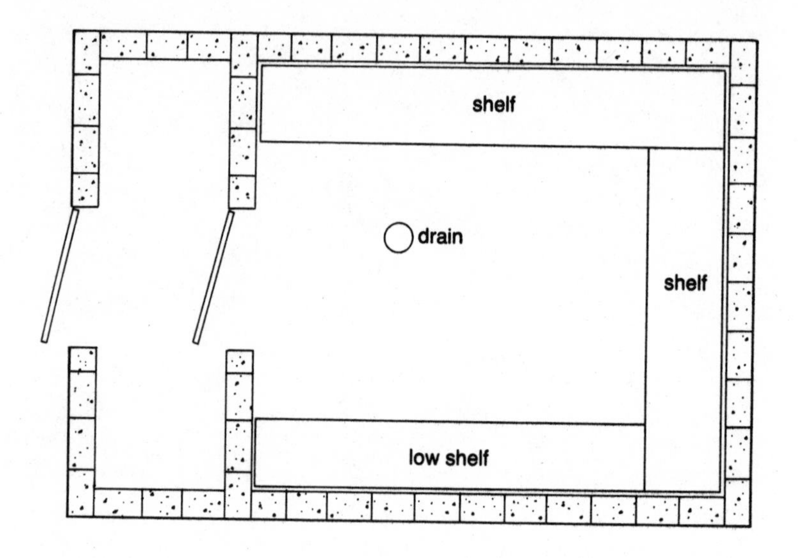


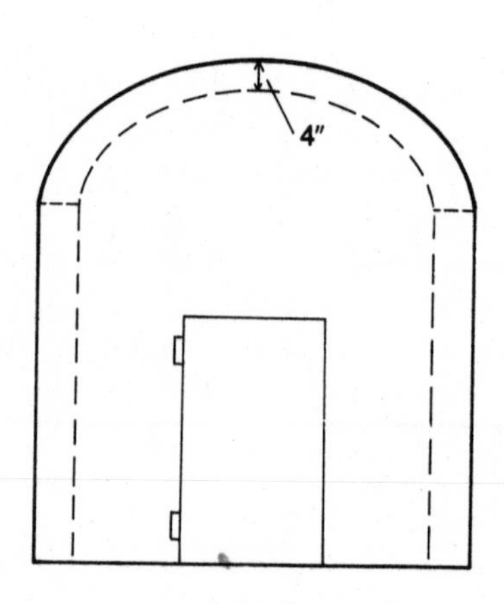

pipes leading to the outside — one 2-inch pipe in the anteroom and a 3½-inch pipe in the main part of the cellar.

Materials and backhoe labor to build the root cellar cost $400. Other labor — block laying and mortar mixing — was supplied free by friends and relatives and, of course, by the Senfts themselves. The backhoe was used to dig the trench for the poured footing, excavate the hole in the hill into which the structure was built, and replace the soil around four sides of the completed block building.

How well does the system work? The owners appear to be quite gratified with the results of their investment in winter security. They keep a thermometer on the shelf and they've noted that winter temperature has remained pretty constant at 38 degrees F. When the outside temperature stayed in the low teens for more than a few days, the cellar temperature dropped to 34 degrees, but never, as far as they know, to freezing. There is only one management problem, and that is a mild one. Water condensing on the ceiling drips onto the potatoes stored below. Since the potatoes are in slatted boxes, though, and the gravel floor affords good drainage, there's been no problem with rotting produce. Mike suggests that one way to avoid this problem would have been to make a somewhat steeper slope in the roof, with ribs running across the ceiling to direct the dripping moisture toward the edges of the storage area.

As we stood in the cool, damp underground hideaway on a hot early fall day, admiring the good workmanship and careful planning that went into its construction, we couldn't help but think how pleasant it would be to come to that same place in the dead of winter, huffing a bit, perhaps, from having climbed through snowdrifts, and fill a basket with the fruits of the summer garden, in a room that would then feel warm and protected from the sharp winds. Stew for supper? Apple fritters? Baked potatoes? The root cellar makes it all possible!

## One That Fizzled

This is the story of a root cellar that didn't work out quite as well as some of the others we visited. We include it here to show that the commonly accepted principles of root cellar construction really do make sense in practice.

When the Moyers (not their real name) built a little shed in their yard to store tools, they decided to include a root cellar too. Avid gardeners, they raise most of their own vegetables. We've seen hanks of fat onions tied to their garage rafters — proof of their gardening expertise.

The Moyers' root cellar is a 5-by-5-foot dirt-floored area under the toolshed. Concrete blocks nine courses high line the walls. The joists supporting the root cellar ceiling (which is the floor of the toolshed) are pine two-by-sixes set 16 inches on center, resting on the top course of blocks. Shelves, made of pine and spruce construction lumber, are 2 feet deep. There are no drains or ventilating pipes. Entrance to the storage area is by way of a trapdoor in the shed floor, with a ladder leading down to the root cellar.

We first noticed the root cellar three years ago when it was being built. When we revisited the place this year, the shelves and joists (which had been made of new lumber) were already rotting — the victims of humidity. Lack of air circulation, we believe, had a lot to do with the short life of the wood, especially soft wood like pine and spruce which are not known for their resistance to decay. Locust or cedar would have given longer service in this damp, stagnant air.

The Moyers used their cellar to store carrots, potatoes, and apples (the three crops most often stored in root cellars, according to our informal survey). They said, though, that to their regret they used the cellar less than they had expected. With the entrance by way of a somewhat rickety ladder (also a victim of the moist, dead air) and the tendency of the toolshed to accumulate boxes of rope, coils of chains, and lengths of scrap wood — usually right on top of the trapdoor — they found it easier to head for the freezer for a package of beans than to kick aside these resource piles to gain access to the carrots below.

Habit, need, and family eating patterns undoubtedly influence the use or disuse of different vegetable storage methods. In this case, the cellar was not sufficiently convenient to encourage the establishment of a new pattern— that of raiding one's own underground for fresh winter vegetables. We could picture the shelves laden with boxes of beets, turnips, cabbages, and kohlrabi, with planter bins of celery, Chinese cabbage, and escarole ranked along the sides on the ground. (The garden at this place is on limestone soil, which is excellent for growing vegetables.) New owners have recently occupied the house. In order to use the root cellar, they will need to replace or shore up the decaying joists and perhaps enlarge the trapdoor, erect storage shelves in the toolshed to keep the clutter off the floor, and add at least one vent pipe to the outside. The space is there. It may yet serve its intended purpose more completely.

# The Cellar Barter Built

Charlie Knier built his root cellar into the slope of a steep northwest-facing hill. In the best tradition of neighborly cooperation, the man next door excavated the hole in the hill with his backhoe and later helped Charlie to lay up the blocks and pour the concrete cap, in trade for help from Charlie on one of his projects. The reciprocal arrangement they had with their neighbor helped the Kniers to hold out-of-pocket costs for their root cellar to about $60, eight years ago.

The eight-foot-square cave is enclosed by nine courses of concrete blocks. Charlie retained the dirt floor, using poured concrete only to form the footing and the roof cap. His experience has been that adequate ventilation and moisture are difficult to maintain in a concrete-floored storage room.

The ceiling of the root cellar is a four-inch-thick slab of poured concrete. Knier made a form for the concrete by propping two carefully levelled four-by-eight-foot sheets of exterior plywood (an exact fit for the

Outside the Knier root cellar.

## Kniers' Root Cellar

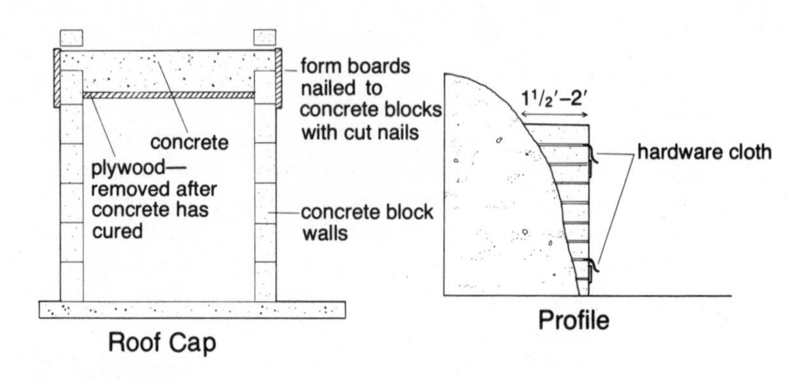


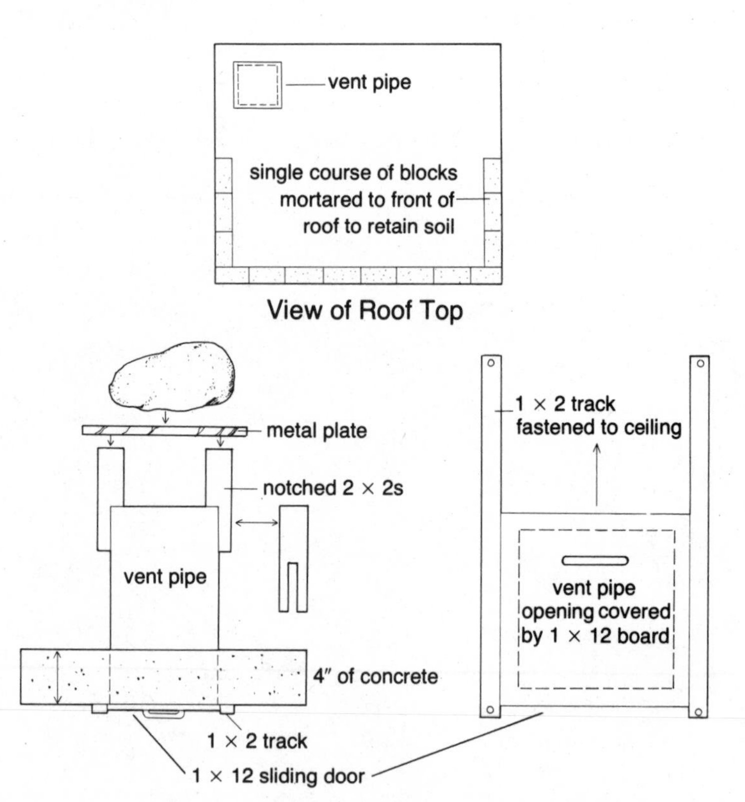

A terra cotta flue liner serves as the vent pipe for the Knier root cellar.

interior space) in place with sturdy framing lumber. The boards that made up the sides of the form were nailed to the top course of blocks with case-hardened nails. (It's a good idea to use goggles when working with these nails because small pieces of the metal sometimes chip off and fly with great force.) Before pouring the concrete, Charlie placed metal reinforcing mesh threaded with reinforcing rods directly on the plywood. He then poured two inches of concrete, jiggled the mesh so that the wet concrete would pass through it, and poured another two inches of concrete on top. Thus the reinforcing metal was embedded in the concrete. The Kniers were careful to stay out of the root cellar while the concrete was still liquid to avoid dislodging a support and collapsing the form and the tons of concrete it held in place. They were also patient enough to allow the roof cap to cure sufficiently before removing the plywood and supporting posts.

The ventilating system was built into the root cellar right from the start. Before pouring the cap, Charlie had set a 3-foot-tall terra cotta flue liner on the northwest corner of the plywood. This ventilating pipe thus became embedded in the ceiling. It's screened on the top to keep out foraging animals and capped by a 16-inch-square metal plate slanted towards the rear and supported by notched two-by-fours to keep out rain. To control the amount of outside air entering the root cellar, Charlie rigged up a neat sliding door in the ceiling — just a board with a handle on it, held in place by two one-by-two-inch tracks nailed to the ceiling over the flue pipe opening, as shown in the diagram.

This arrangement has been adequate as long as Charlie remembers to open the door wide during periods of prolonged rain. If it were necessary, ventilation could be improved by installing a screened, flap-covered hole in the door.

During mild weather, when the outdoor temperature is considerably higher than the temperature of the root cellar, moisture condenses in droplets on the ceiling. Any moisture that drips to the floor is absorbed by the gravelled dirt surface. Charlie has installed a drainpipe in the floor, though, to carry off excess moisture. The two-inch-diameter black plastic pipe he used is perforated on the top but not on the bottom; therefore it can collect water and carry it off to a spot several feet away from the cellar.

The fir framing lumber used for the original shelves has lasted about four years, but as the photo shows, dampness has taken its toll. Charlie intends to replace the softwood shelving with storage racks made of locust.

On the April day when we visited, there were still apples and potatoes in the root cellar, in usable condition. Apples make excellent sauce after storage, Charlie told us; they need only a little cooking and hardly any sweetening. He's had no problem storing apples and potatoes together — has done it for years, without any undue premature sprouting of his stored potatoes. An accomplished gardener, Charlie stashes away all the extra root vegetables from his garden at harvesttime in the fall.

The cellar's average temperature, year round, is about 47 degrees F (near Harrisburg, Pennsylvania). In summer it climbs to 50 degrees, but that's tops. The lowest winter temperature was 40 degrees. The steep hill and surrounding earth plus shade from trees help to maintain this nearly constant cool temperature.

As the diagram shows, a single course of soil-retaining concrete blocks is cemented in place around the edge of the roof. The top of the roof was covered with one foot of soil when the concrete had cured. This soil has now settled and it is time to add more to keep the roof well

The Knier root cellar has storage shelves on each side of the central aisle, and a gravel-covered dirt floor.

The insulated door of the Knier cellar with hasp.
Note the light to the left of the door.

insulated. It's important, too, to clear off all new growth of trees and shrubs from the roof of the root cellar because their roots can damage the mortar.

The Kniers' root cellar is further insulated by a layer of two-inch Styrofoam on the door. The exposed front wall had been fitted with two-inch-thick batts of fiberglass, but the fiberglass absorbed so much moisture that it lost its insulating value. Charlie's next project will involve either replacing the fiberglass with two-inch Styrofoam, or building an anteroom in front of the present door.

Luckily, Charlie had the foresight when building the block walls of the cellar, to cement flaps of ¼-inch hardware cloth between courses of the end blocks on the front face of the cellar. These ties will enable him to attach the small front addition by cementing them between courses in the new wall. Without some method of bonding the old and new structures together, the small additional anteroom would be unstable.

Charlie generously noted several other fine points of root cellaring that he's learned from experience:

*Carrots and beets:* These keep best on a low shelf away from ceiling moisture if droplets of condensation are dripping from the ceiling.

*Mice:* "Although no one has ever found out how mice get into a presumably secure root cellar," Charlie told us, "they do get in." He sometimes finds it necessary to nail wire screening around the crates in which he keeps apples. Another, easier trick he suggested is to leave some "sacrifice" fruit — poorly shaped or already bruised or partly spoiled — accessible for the varmints to eat in hopes they'll ignore the sound food you're saving for your family.

*Door hardware:* A hasp and staple makes a good door closure. If you mount the staple on the door and the hasp on the wall, you'll be able to cover the lock, if you use one, with a flap of leather or rubber to protect it from the weather. Set hinges into the concrete with case-hardened nails.

*Wiring:* "Be sure to use outside wiring in the root cellar," Charlie cautioned us. "Regular indoor wiring will corrode in all that dampness. And if you do wire in a light bulb, use a porcelain socket and install an outside or in-the-house switch for convenience."

## Truck Storage

This root cellar makes appropriate use of a discarded truck body — a nice trick, and an almost "instant" installation. W. N. Woodzell of Earlysville, Virginia, found that his root cellar space came ready-made when he bought an old street-delivery milk truck from a junkyard. The truck has double doors on the back and — as luck would have it — the body of the vehicle is insulated and lined with aluminum. The aluminum lining makes a clean, light, rodent-proof interior for the cellar. We wouldn't suggest buying aluminum to line a regular root cellar, since it requires a great deal of fossil fuel to produce. However, when the stuff already exists, but is in disuse, the only thing to do is to make good use

## Woodzells' Truck-Body Root Cellar

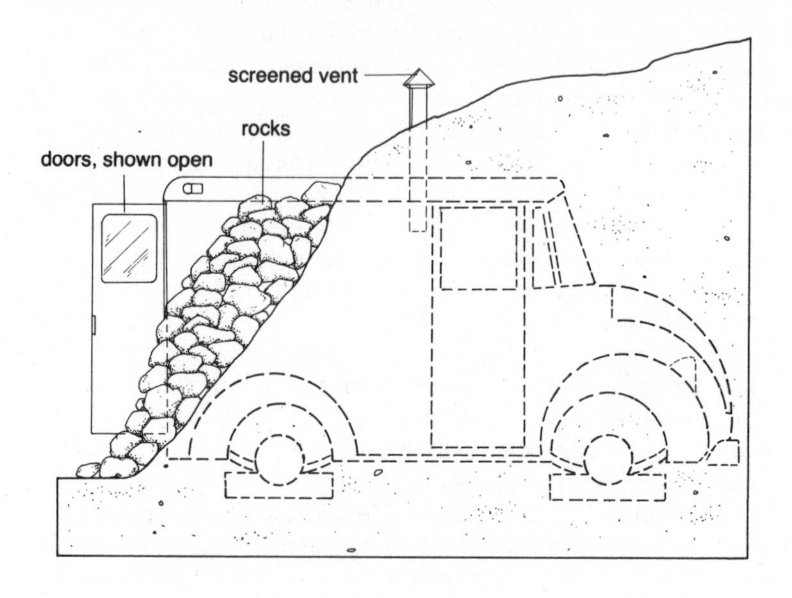


of it and enjoy it. That's what is fun about recycling — you often wind up with something even nicer than your conscience would have let you buy.

Woodzell hired a backhoe to dig out a hole in the side of a hill on his place. The site of the root cellar is a wooded slope so there is plenty of shade to keep the storage enclosure cool. A vent cut in the truck roof is fitted with a pipe screened at the top to keep out animal and insect intruders. Woodzell simply drove the truck up into the hole in the hill, removed the wheels and then backfilled with soil. Rocks piled up at the edges of the exposed sides further protect and insulate the truck body. How does it work? "I did this six years ago," exults Woodzell, "and it's still a great root cellar!"

Winter temperature averages around 40 degrees F. It goes up to 55 to 60 degrees in summer. The Woodzells store turnips, white and sweet potatoes, and squash in their buried truck. They say they've had no problems with their vegetable storage. And when they needed a shallow well pump they installed it in the front of the root cellar — a nifty way to provide good housing for the pump and dampness for the cellar. (You'll note that the Woodzells keep squash and sweet potatoes, which ordinarily require warm dry conditions, in their truck larder. Every authority we've read warns that these vegetables will rot in cold damp places and our own experience bears this out, but we think it's important too to let you know what people are really doing. Some folks do keep onions in cold moist cellars and canned goods in damp places and get away with it. Much depends on humidity level and amount of ventilation — and, of course, on length of storage. Some vegetables will keep for a few weeks in the fall under less-than-ideal conditions but they wouldn't last for the winter.)

## A Small Root Cellar Is a Big Improvement

Like many other accomplished gardeners, John and Kathryn Hill were well settled in their new house when they realized that they needed a good place for root vegetable storage. And, as Kathryn puts it, "Each summer we find ourselves trying to improve something around the house besides running the garden. We put this in several years ago and wouldn't know what to do without it." The small but solid root cellar the Hills built ranks as a most useful improvement.

The root cellar is situated right next to the basement, at the bottom of the exterior steps leading to the basement door. Thus it can be reached both from the outside and from the inside through the basement door leading to the outside steps. In the coldest winter weather the Hills close over the exterior stairwell with boards laid across at ground level, but in the fall the exterior stairway is handy for bringing in the harvest.

John Hill dug the hole for the five-by-five-by-six-foot root cellar himself. He started on a Friday evening and finished the job late the next day. Since the soil had recently been backfilled around the new house basement, it was relatively loose and not too hard to dig.

Then John cut through the concrete block house foundation to make a doorway leading into the adjoining root cellar space. He also installed a drain leading from the root cellar and connected it to the cellar drain.

The next step was to make the plywood forms for the poured concrete walls and ceiling. This was the trickiest part of the whole operation. The plywood forms started to bulge during the pour so some quick footwork was needed to shove in a few more two-by-four braces in the space between the forms. It all worked out very well. The walls and ceiling are four inches thick. For reinforcement in the ceiling the Hills cleverly used the inner steel frame and wires from an old upright piano.

There is no air intake vent, but the four-inch exhaust pipe rises to a height of three feet above the ground to keep it visible and prevent

## Hills' Root Cellar

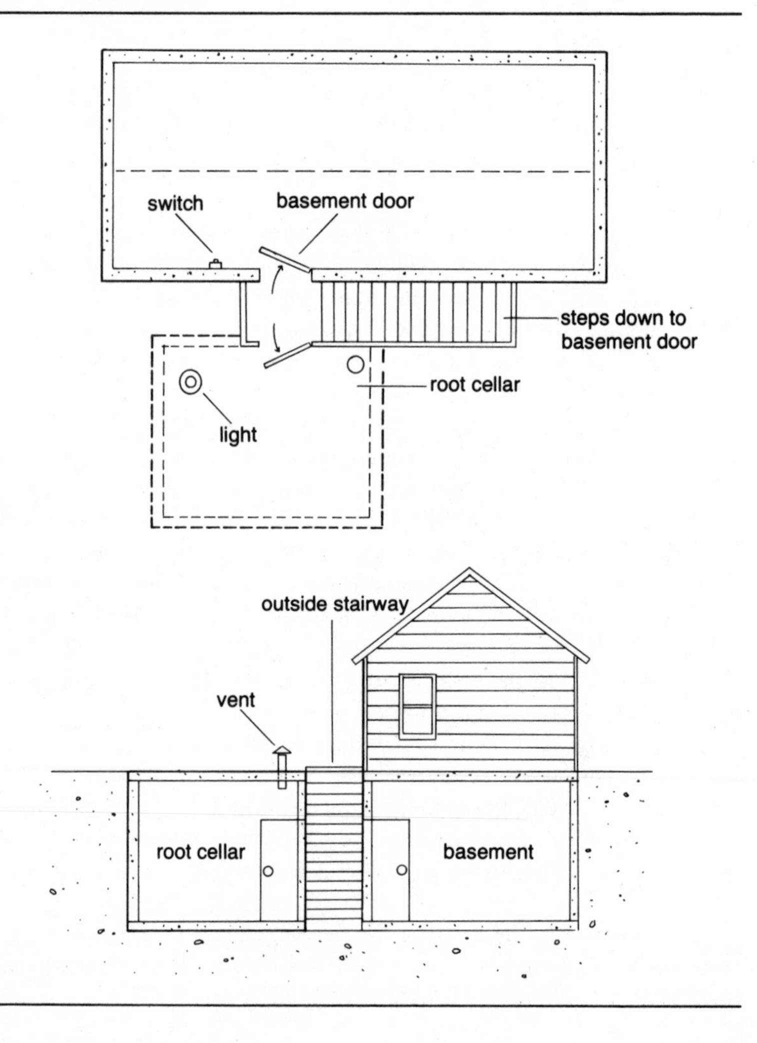

Lugs of homegrown potatoes and a bin of carrots stored in the Hills' root cellar.

accidental tripping. A nearby section of wrought iron fencing that runs along the edge of the basement stairway also helps to keep traffic away from the air vent.

The Hills keep potatoes, carrots, celery, and cabbage in their root cellar. Dahlia, gladiolus, and canna bulbs are stored there, too. Instead of shelves, they stack crates of potatoes along the wall. Carrots kept in sand were still in fine shape in March. The cellar ceiling sometimes gets frosty in very cold weather but the produce does not freeze. Although they don't keep a thermometer in the cellar, the Hills leave an open jar of water in the room. If it were to begin to form an ice skin they could open the cellar door briefly to warm it up.

## Another Enlarged-Basement Cellar

Like the Hills, Mr. and Mrs. Phil Cecala of Ingleside, Illinois, dug a root cellar adjoining their house basement, accessible through the basement door. Six men worked on the excavation off and on for a month. "I tell you, it's a lot of work," Phil exclaims, "but I would do it over again and again. You will never be sorry!"

The Cecalas poured the six-inch-thick concrete walls and eight-foot-high arched ceiling of their six-by-ten-foot root cellar. They wisely included a low-placed air intake pipe and a high-up exhaust vent. The form used for pouring the arched concrete ceiling was made of tapered two-by-eights as described in chapter 16. Cecala reinforced the ceiling concrete with steel rods. "Much stronger than mesh," he said. He brushed on a coat of tar over the outside of the concrete before it was completely cured. (Let the tar dry thoroughly before backfilling, though, or the earth and rocks will scrape it off.) The cellar is topped by two feet of earth.

The Cecalas keep carrots, celery, celeriac, potatoes, beets, and homemade wine and beer in their capacious cellar. They also have an open-dug well with a hand pump in the cellar, with an interlocking four-inch drain tile going down to below water level. When electricity goes out, they're never without water. A nifty arrangement, we'd say.

## An Underground Root Cellar

When Arthur Schroeder of Big Rapids, Michigan, built his underground root cellar, he followed Michigan State University plans that specified a poured concrete wall. Schroeder substituted concrete block walls and had extra concrete poured into the hollows of the blocks when the ceiling cap was poured. His original bulkhead door was installed at a low angle to the ground but froze over in winter so last summer he changed the opening to an upright door.

## Schroeders' Root Cellar

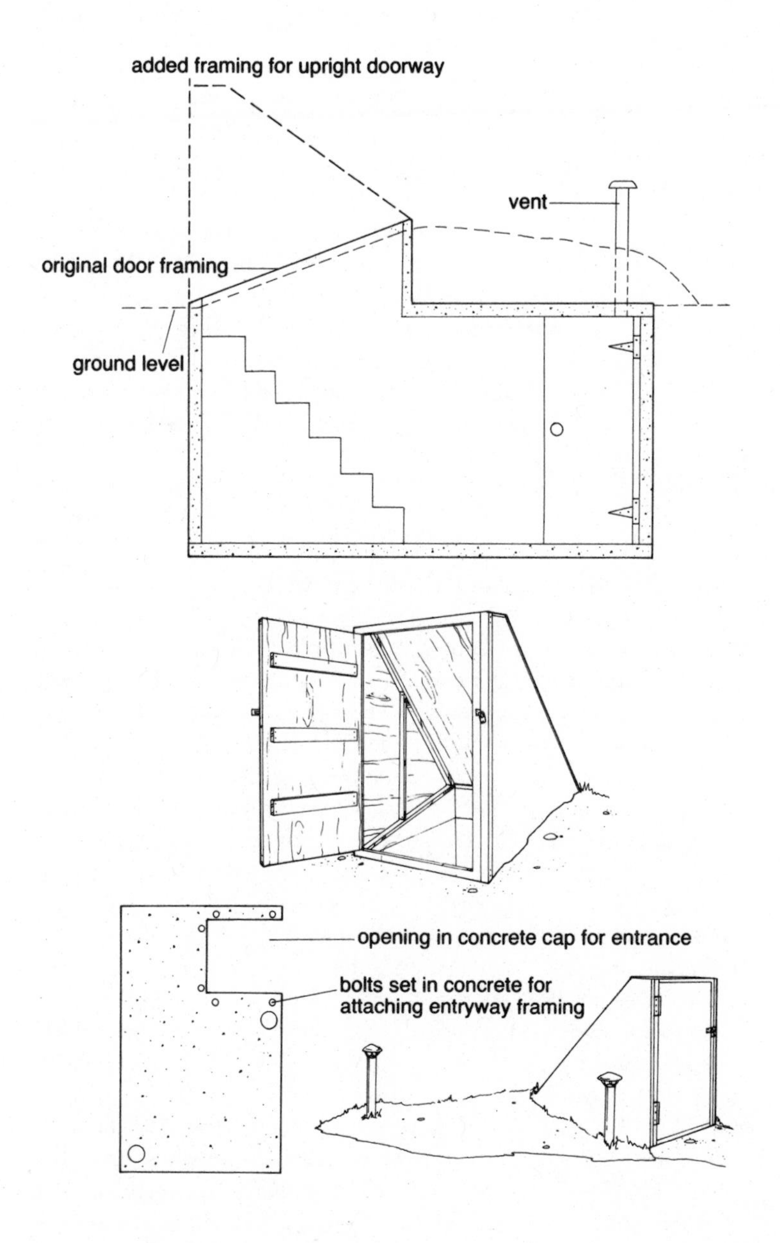

After covering the cellar with earth, Schroeder planted crown vetch on the mound to keep the soil in place.

"Temperatures in my cellar range from 50 to 54 degrees F summer and winter," Schroeder reports. "It will maintain this range if both doors are kept tightly closed. Last year we were still using potatoes from our root cellar until the last of June. They remained firm and didn't sprout at all. We also store glad bulbs, onions, and home-canned fruit there, since our mobile home has little room for canned goods.

"I made another adjustment in addition to rebuilding the entryway. Twice last summer I found dead bluebirds in the cellar. They had flown in through the open vent stack and couldn't get out. I've screened the vents now so as not to lead any more bluebirds astray.

"Incidentally," Schroeder adds, "the cellar makes an excellent tornado shelter, as long as you take along a few blankets to ward off the chill. I keep several folding chairs in my root cellar in case I need it as refuge from a storm."

## Partitioning a Ranch House Basement

When Bill and Sarah Graham moved into their ranch house in a suburban development, the basement was finished but not divided into rooms. There was a workshop at one end of the basement, a fireplace and furnace corner at the opposite end, and a big open space in between. Out in the yard, there was a good big garden patch started by the former owners, a grape arbor, and a row of dwarf fruit trees that would soon start bearing. The Grahams are both enthusiastic gardeners and creative cooks who like to use good fresh well-grown whole foods in their cookery. When they looked at that big open space in their basement, it didn't take them long to decide that they had a good spot for that cold cellar they'd always wanted to have.

And so they got busy, that first winter, and partitioned the basement into rooms. On the north wall, near the outside basement door, they enclosed an 8-by-8-foot room with two-by-four studs. They faced the outside of the studs with woodgrain panelling. Then they inserted sheets of 2-inch Styrofoam between the studs. They insulated the door and ceiling with Styrofoam, too, and they panelled the walls inside the cold cellar with ⅜-inch exterior plywood. The floor is concrete and the high screened window admits cold air. There is a floor drain just outside the cold cellar in the hallway leading from the back door.

## Grahams' Root Cellar

Original Basement

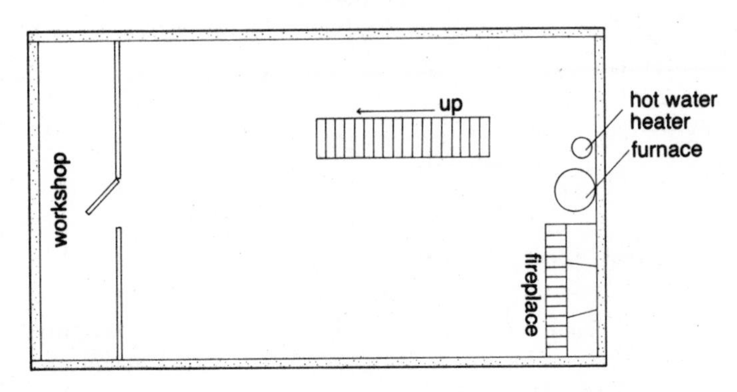


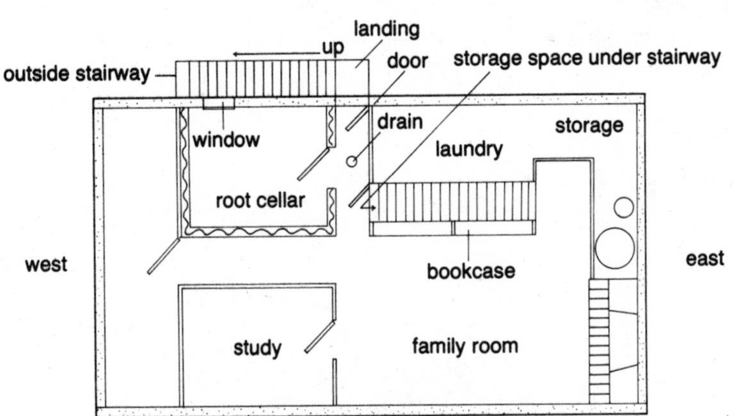


After Partitioning

Bill built free-standing shelves of rough oak lumber to line three sides of the room. The Grahams keep apples in crates, carrots, beets, and turnips in leaves and sawdust, Chinese cabbage and escarole in tubs on the floor, Ballhead cabbage on the slatted shelves, and pears, grapes, and potatoes (the latter purchased from a local grower). In the cold hallway leading to the back door, where the air is not as damp as it is in the root cellar, they hang braids of garlic and net bags and braids of onions. They store their squash on a high shelf in the workshop, which has no heating ducts but stays around 50 degrees F if the door is left slightly ajar.

The basement renovation took careful planning and — to get the whole project done — a good many hours of weekend and evening work, but Bill and Sarah agree that dividing up the space has multiplied its usefulness. And they continue to develop more garden corners on their suburban lot so they can grow even more root cellar vegetables.

## A Stone Root Cellar

A fine stone root cellar like this one of Bob and Autumn Wooling's is almost too handsome to bury in the ground. Enough of the stone facing remains visible, though, to give them a well-earned glow of satisfaction when they go to raid their cellar for some of their many homegrown foods.

### Woolings' Root Cellar

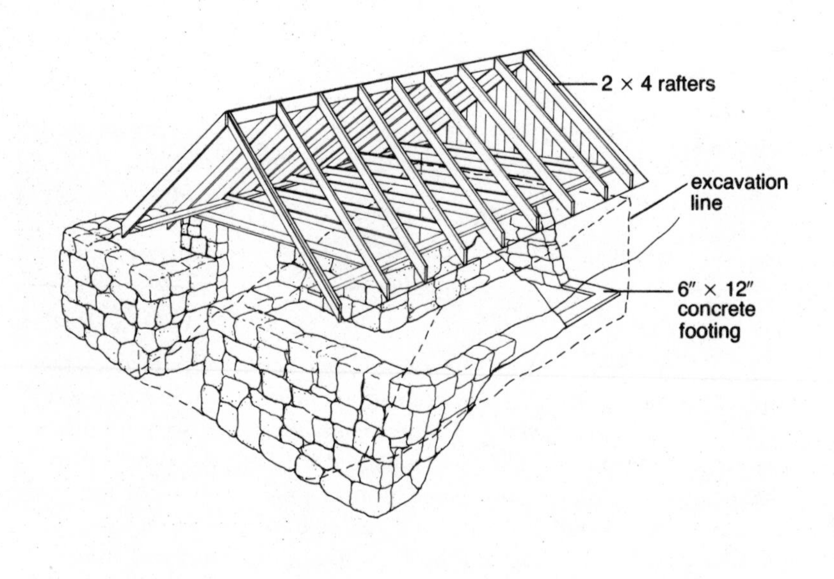

"With the help of friends, we dug a 12-by-15-foot hole in the hillside on the site of an old caved-in root cellar," Bob recalls. "We got the digging done in the usual way," Autumn adds. "We told everyone we wanted to dig a root cellar on a certain day. We cooked lots of food and everyone dug a little bit. By evening it was all dug.

"Then," continues Bob, "I poured a footer 6 inches deep and 12 inches wide. We spent a long time building up the rocks. The cellar took a year and a half to build. There were some nice square rocks from an old house chimney and we found others around the place. The only one that we really had to search for was the long thin door lintel. We didn't use forms. I laid up random rubble walls to a height of 6 feet, leaving a 2-foot opening for the door. The average wall thickness is 10 inches. I had noticed that the front wall of the original (caved-in) root cellar was a foot wider than the back wall. I figured that the old-timers knew what they were doing so I built mine the same way.

"The top sills are oak two-by-eights. To attach the sills to the rock walls, I drove one-inch galvanized roofing nails into the sill pieces and then worked very rich mortar onto the board and between the nails. At the same time, I put a layer of this mortar on top of the stones. Then I turned the mortared nail side of the sill onto the stones, and left it alone until it set up. The roof is framed and top-floored with oak two-by-fours. I built in two fiberglass-insulated doors for extra protection against

The Woolings' root cellar—ready for the roof.

## Woolings' Root Cellar

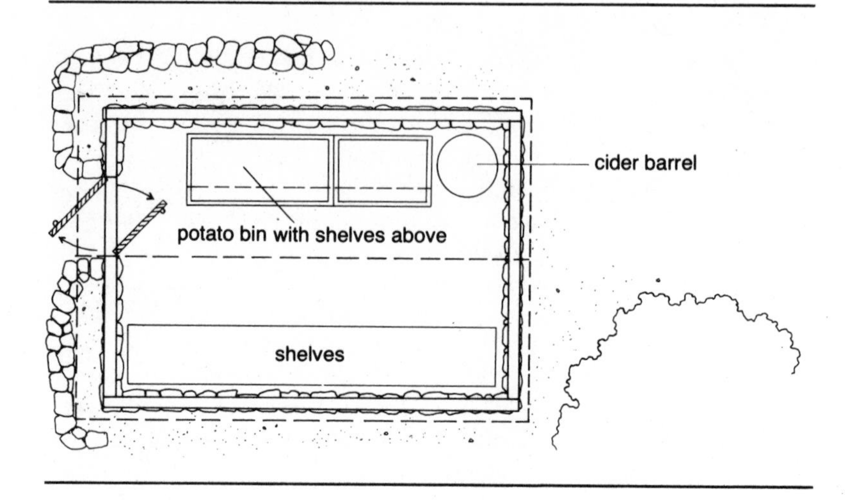


weather extremes. There's a small vent shaft going through the ceiling and roof.

"The ceiling is insulated with six inches of fiberglass sealed between aluminum offset press plates (obtained from our local newspaper). I also attached Celotex to the exposed ceiling surface inside the root cellar, but this was not a good idea. The Celotex absorbs too much moisture."

The Woolings' root cellar is built into a south-facing hill, which is apparently the custom in their area. To insulate the cellar from the warming sun, they've laid up loose rock walls in front of the exposed walls and filled in behind them with two feet of dirt, so that there is a double wall on the south side. In addition, they planted forsythia bushes in front of the wall, both for decoration and further insulation.

"The weather in our part of West Virginia gets as cold as New England in winter because of the mountains," Bob relates. "But in summer it will stay in the 90s because of our more southern latitude. The cellar maintains a steady 65 degrees F in summer and stays above 33 degrees in winter, averaging about 34 degrees." Autumn adds, "Our cellar temperature is incredibly even. In the fall it drops to 34 degrees and just stays there. Only if the outside temperature is below zero for a week do I put a coal-oil lamp in the middle of the floor. Last winter was very cold here and the front inside wall of our cellar got white with frost but nothing in the cellar froze.

The Woolings' structure as it nears completion.

## Woolings' Root Cellar

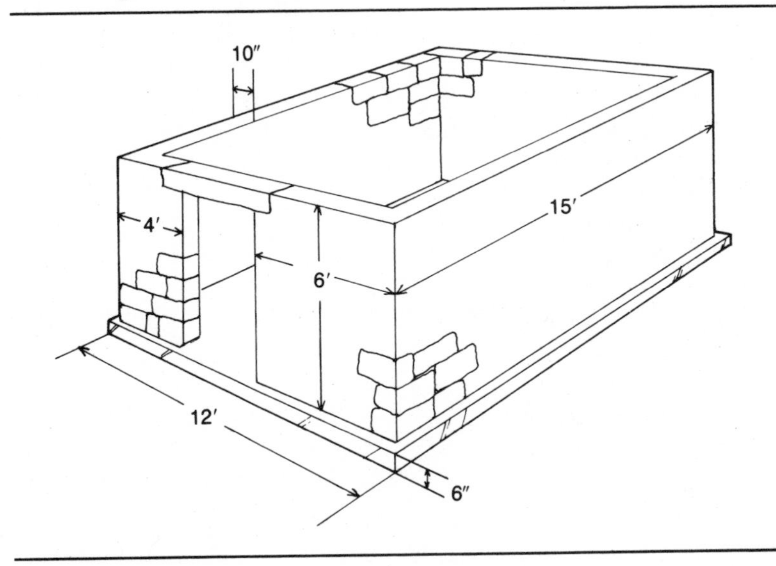

"We keep sawdust on the floor. In the summer we leave the doors open for several days to really air the cellar out. It's handy to have a heavy screen door that opens outward to keep chickens and dogs out in summer but lets the sun shine in to kill molds and fungus. We've learned from the mistakes we noticed in other root cellars. Many people make their cellars too small, make the shelves too wide, and the walls too thin. One interesting thing people do around here, and it often works very well, is dig their root cellar into a hill and then build a smokehouse over it."

Drainage is important too. According to Bob, "People make a big mistake by not keeping a good roof on their cellars. There should be gutters around the roof and drains around the footing. The only reason the original old root cellar here on our place collapsed was because water from a bad roof got behind the walls and froze."

If you were to write an advertisement for the good life, you couldn't improve much on the Woolings' list of the provisions that they keep in their root cellar. When we asked Autumn what they store in their cellar,

## West Virginia Smokehouse/Root Cellar

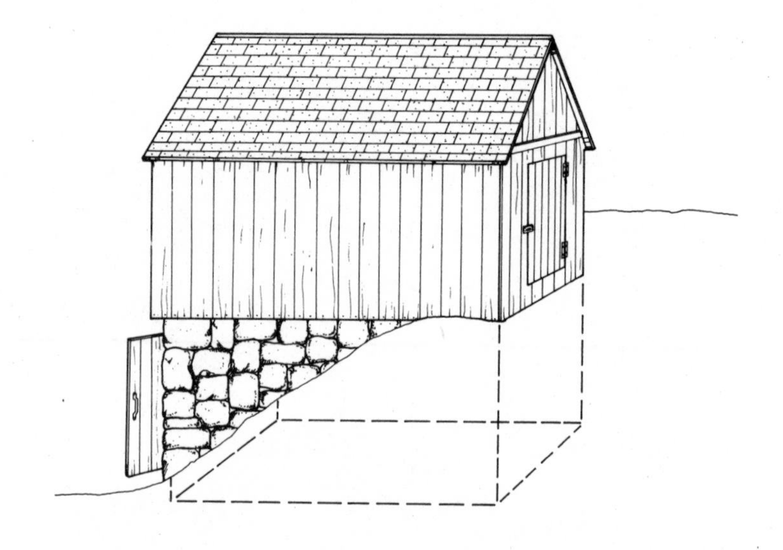

she replied, "Well, I guess strawberry jam in June is about the first thing, then peach jam, rhubarb, gooseberry, currant jelly, wild plum butter, blackberries, raspberries, then applesauce from yellow transparents. By August it's tomatoes, sauerkraut, all kinds of pickles, corn sallet, and relishes. We have an 'apple butter tree' that makes the best apple butter you ever tasted — I make a large amount of that. Then more tomatoes (about 150 quarts). We have a wide shelf over our potato bins and we put pulled-out-with-roots cabbage up there and sometimes pears and apples. In the fall we make cider too. Then we dig the potatoes and beets and carrots and keep them in sand in bins. Then later in the fall there's butchering and I can lots of sausage and lard. That's all till February, when I can maple syrup from our trees and store it too in the cellar. I freeze a few vegetables but mostly depend on the root cellar."

# SECTION SIX

## Recipes

# 18

# Cooking Sturdy Keepers

Food is our common property, the body of the world, our eating of the world, our treasure of change and transformation, sustenance and continuation.

Edward Espe Brown
*Tassajara Cooking*

On any given winter day, gardeners all across the land who've prepared for winter will be heading for their root cellars to choose the makings of the evening dinner. If you're one of them, perhaps you'll be thinking, as you fill your basket or stuff your pockets, about how you'll fix the vegetables for supper. It's easy to get into a rut and serve the same boiled turnips or carrot casserole regularly. We all need new ideas, from time to time, for imaginative ways to use the sturdy good keepers from our winter stockpile. Here are some of our family's and friends' favorite ways of using root cellar food. We hope you'll find among them some that will become your favorites, too. (You'll notice that in most cases we've left out the salt. Most of us eat too much of it. If you want to add some at the table, that's up to you.)

# Salads

### *Apple-Cheese Salad*

3 large eating apples, diced
1 cup alfalfa sprouts
1 cup diced cheese — Longhorn, Cheddar or Muenster are good
1 cup diced thick ends of Chinese cabbage leaves
⅓-½ cup mayonnaise (see page 280)
2 tablespoons lemon juice

Toss ingredients together. Serve plain or on a bed of escarole greens.

Serves four.

### *Sunshine Pear Salad*

1 cup orange sections, cut from peeled oranges
6 fresh pears, peeled
1 avocado, peeled
1 pound cottage cheese

Arrange fruit around cottage cheese, either on individual plates or on a large serving plate. This makes a meal with pumpkin or sunflower seeds and hot bran or cornmeal muffins.

Serves three for a main dish, six for a side dish.

### *Belgian Endive Salad*

4 heads Belgian endive, washed and chopped
1 onion, sliced thin
1 large carrot, shaved into thin slices with vegetable peeler
1 avocado, diced
1 cup alfalfa sprouts
3 medium Jerusalem artichoke knobs, sliced
1/4 cup salad oil
1/4 teaspoon oregano, dried
freshly ground black pepper
1/8 cup vinegar

Toss vegetables in oil. Add herbs and vinegar and toss again.

Serves four.

### *Escarole and Other Good Things*

1 head escarole, washed, whirled dry, and torn in pieces
1 onion, sliced thin
2 tomatoes, sliced
1 clove garlic, crushed (optional)
2 small kohlrabis, sliced
4 sprigs parsley, minced
2 hard-boiled eggs, sliced
1/4 cup sunflower seeds
1/4 cup oil
3 tablespoons vinegar

Toss all ingredients with oil. Add vinegar and toss again.

Serves four.

### *Midwinter Salad*

1 small head Chinese cabbage, sliced thin
1 winter radish, sliced paper thin (use slaw board)
1 onion, chopped
¼ cup ripe olives, sliced
1 clove garlic, crushed
¼ cup oil
3 tablespoons lemon juice
⅔ cup broken whole wheat pretzels

Toss vegetables with oil. Add lemon juice. Add pretzels and toss again. Or serve with blue cheese dressing.

Serves four.

### *Blue Cheese Dressing*

¼ cup crumbled blue cheese
½ cup yogurt
½ cup mayonnaise

Mix all ingredients.

### *Hearty Herring Salad*

2 cups shredded red cabbage
2 washed, unpeeled eating apples, diced
1 cup pickled herring, chopped
1 cup cooked, diced beets
½ teaspoon celery seed or dill seed
1 tablespoon honey
¼ cup salad oil
3 tablespoons lemon juice

Toss all ingredients with honey, oil, and lemon juice.

Serves four.

### *Special Potato Salad*

4 cooked potatoes, peeled
½ cup cubed cooked celeriac, peeled
1 onion, diced
2 hard-boiled eggs, sliced
½ teaspoon dried dill seed or leaves

Toss above ingredients with mayonnaise (see page 280). Garnish with ground paprika and parsley.

Serves four.

### *Country Cole Slaw*

3 cups shredded cabbage
2 apples, washed and diced unpeeled
¼ cup chopped pickles — watermelon, cucumber, or other
¼ cup yogurt
¼ cup mayonnaise
3 tablespoons pickle juice
apple wedges, sliced thin

Mix all ingredients and garnish the salad with thin-sliced apple wedges.

Serves six to eight.

### *Year-Round Carrot Salad*

4 cups grated carrots
1 cup raisins
1 cup mayonnaise
1 cup drained, unsweetened crushed pineapple
1 tablespoon lemon juice

Mix and chill until serving. Last year I made this in May from carrots kept in sawdust all winter.

Serves six.

# Soups

### *Pea Soup Plus*

1 pound dry split peas
2 cups cubed winter squash
2 stalks celery
1 bay leaf
3 quarts water or stock
1 onion, sliced and sautéed until transparent
2–3 tomatoes, fresh or canned

Simmer peas, squash, and celery with bay leaf in hot water or stock until tender, about 30 minutes. Add sautéed onion and tomatoes. Heat thoroughly. Remove bay leaf.

Serves four to six.

### *Pumpkin Bisque*

1 onion, chopped
2 leafy tops of celery stalks, chopped
2 tablespoons butter
2 cups milk
2½ cups mashed cooked pumpkin (or squash)
2 cups chicken stock
pinch of cinnamon
pinch of cloves

Sauté onion and chopped celery leaves in butter until tender. Gradually add milk to mashed cooked pumpkin and stir till smooth. Then stir in chicken stock. Add spices. Heat and serve.

Serves three to four.

### *Mock Oyster Stew*

1 onion, diced
1 tablespoon oil
4 cups milk
2 cups cooked, sliced salsify root
½ cup cooked chopped celery or celeriac
chopped parsley or chives
pepper

Sauté onion in oil. Combine all ingredients, except parsley and pepper, and heat slowly. Garnish with chopped parsley or chives. Pass the freshly ground pepper.

Serves four.

### *Farmer's Winter Soup*

2 carrots
1 large onion
3 small beets
2 potatoes
1 turnip
2 stalks celery, or ½ cup cubed celeriac
2 cups chopped cabbage
1 winter radish, sliced in rounds
2 quarts water
1 bay leaf
1 clove garlic speared on a toothpick (fish it out when soup is done)
3 medium tomatoes, chopped (or 1 cup canned tomatoes)
1 quart beef stock
yogurt

Cut vegetables in small pieces and drop into simmering water with bay leaf and garlic. Cook until tender, about 20 minutes if diced small. Add tomatoes and stock when cooking is complete. Serve soup with yogurt spooned on top. Flavor is best the day after making, although some vitamins are lost in reheating.

Serves six to eight.

### *Cream of Cabbage Soup*

3 cups shredded cabbage
1 medium onion, diced
2 tablespoons oil
1 cup water
2 potatoes, peeled and diced small
3 cups chicken broth
1 cup milk
paprika
3 tablespoons whole wheat flour (optional)

Sauté cabbage and onion in oil for 5 minutes. Add water and potatoes and simmer for 15 minutes. Add hot broth and hot milk. Serve with a dusting of paprika. If you like a thick soup, heat broth separately and then slowly stir into whole wheat flour before adding it to the vegetables. Add the hot milk.

Serves three to four.

### *Leek Soup*

2 cups sliced leeks
2 stalks celery, sliced across
5 medium potatoes, diced
1 cup water
2 quarts hot milk
handful of minced fresh parsley, or 1 tablespoon dried parsley flakes

Simmer vegetables in water until tender, about 20 minutes. Add hot milk. Garnish with parsley. Comforting on a stormy night, especially with fresh homemade bread.

Serves four.

### *Kale Soup*

½ cup barley
½ celeriac root, peeled
2 carrots, sliced
1 large onion, sliced
1 turnip, diced
3 quarts meat stock — lamb, beef, or chicken
3 cups kale, cut into small pieces

Simmer barley and root vegetables, including onion, in stock until almost tender. Add kale and cook for five to eight minutes longer.

Serves four to six.

### *Onion Soup*

2 very large onions, sliced
4 tablespoons oil or butter
2 quarts beef stock
pinch of powdered cloves
toasted bread
grated cheese

Sauté onions in oil or butter until nearly tender but not brown. Add hot stock and cloves and heat gently for 15 minutes. Serve with toasted bread and grated cheese. It's customary to float a piece of crisp toast in each soup bowl.

Serves four to six.

# Main Dishes

### *Patchwork Farm Chinese Dinner*

½ cup sliced Jerusalem artichokes
2–3 cups mung bean sprouts
½ cup chopped sweet peppers
1 onion
1½ cups chopped Chinese cabbage
1 cup green soybeans, cooked
3 tablespoons oil
4 cups water
2 cups brown rice
soy sauce
grated Parmesan cheese

Sauté vegetables quickly in oil, for 8–10 minutes. Serve with brown rice, made by bringing water to a boil, adding raw brown rice and simmering 45 minutes, adding water as necessary toward the end.

Serve with soy sauce (tamari is good) and grated Parmesan cheese. For some reason, green tomato pickles go especially well with this dish.

Serves four, with rice left over.

### *Garden Spaghetti*

1 vegetable spaghetti squash
12 slices nitrate-free bacon
2 cups grated Cheddar cheese
½ cup minced parsley

Steam spaghetti squash in covered pan for one hour. Meanwhile, fry bacon until crisp. Drain, and crumble into small pieces. Slice squash open and fork out seeds. Then tease out pulp with large fork. Top with grated cheese and crumbled bacon. Garnish with parsley.

Serves four.

### *Pumpkin Meatloaf*

1 small pumpkin
hot water
2 slices stale bread or ⅔ cup oatmeal
¼ cup milk
1½ pounds ground beef or lamb
pinch of thyme
3 tablespoons ketchup

Slice top from pumpkin and scoop out seeds and stringy pulp. Pour one inch of hot water into a baking pan and set pumpkin in hot water. Cover pan to hold in steam. Bake pumpkin in 400-degree-F oven for 45 minutes. Soak bread or oatmeal in milk and add remaining ingredients when bread has softened. Then pack meatloaf mixture inside pumpkin and bake at 350 degrees F for 50–60 minutes.

Serves six.

### *Leek Omelet*

1 cup leeks, chopped fine
1 tablespoon oil or butter
5 eggs
3 tablespoons milk
grated Cheddar cheese
toasted bread cubes

Gently sauté leeks in oil or butter for ten minutes. Keep hot while you make omelet. Whisk eggs and milk together with wire whisk or two-tined fork. Pour egg mixture into hot oiled pan. As edges set, lift them carefully with spatula so that uncooked mixture can run under and cook. When omelet is barely set, remove to hot pan, spread leeks over half, and fold other half on top. Serve grated Cheddar cheese and toasted bread cubes for garnish.

Serves three.

### *Ratatouille — a Garden Medley*

3 medium onions, sliced
2 sweet peppers, chopped
⅓ cup olive oil
1 large firm eggplant
4 medium tomatoes, chopped
3–6 cloves garlic, crushed (optional!)

Sauté onions and peppers in part of the oil. When nearly tender but not brown, add other vegetables and heat briefly. Then pour into shallow baking pan and bake at 350 degrees F for 40–60 minutes, adding remaining olive oil just before putting vegetables in the oven. Excellent with rice, and good as a cold spread, too.

Serves four.

### *Apple Stuffing* (or stuff'ning, as our children would say)

2 apples, peeled and diced
2 onions, diced
2 stalks celery
10 cups whole wheat bread cubes
pinch of thyme
1 egg
½ cup milk

Mix apples and vegetables with bread cubes and thyme. Stir egg and milk together and toss into bread mixture. Stuff into turkey or chicken, or bake separately in greased casserole dish at 325 degrees F for 50–60 minutes.

Enough for a 12-pound turkey.

### *Fall Menu from the Garden*

This one couldn't be easier:

4 ripe tomatoes, sliced
4 medium-size sweet potatoes, unpeeled, sliced in rounds and steamed for 30 minutes
2 cups green soybeans, steamed in pods for 8 minutes and popped out of pods

Divide vegetables among four plates. Ring dinner bell. That's it!

Serves four.

## Vegetables

### *Tangy Oranged Beets*

½ cup vinegar (homemade wine that's turned to vinegar is excellent here)
½ cup liquid from cooked beets
¼ cup honey
1 tablespoon frozen orange juice concentrate
2 tablespoons plus 2 teaspoons arrowroot flour or cornstarch
1 big Long Season beet or 3 medium Detroit Red or other conventional-size beets — cooked, peeled, and sliced
1 tablespoon grated orange rind

Mix liquid ingredients.

Blend a small amount of liquid into the arrowroot flour. When smooth, slowly stir in remaining liquid. Heat, stirring frequently, over low heat until mixture thickens. Add beets and grated orange rind.

Serves four to six.

### *Baked Beets*

Parboil small beets to heat them through. Then bake them in a 350-degree-F oven for 50–60 minutes.

### *Crunchy Lunch*

2 or 3 stalks from a head of Chinese cabbage
¼ cup chunky peanut butter (or plain if preferred)
3 tablespoons raisins
¼ cup alfalfa sprouts

Snap stalks from head of Chinese cabbage. Spread peanut butter on thick lower end of stalks. Poke raisins into peanut butter. Sprinkle alfalfa sprouts on top. Cut off leafy upper half of Chinese cabbage stalk and press it on top of the thicker filled lower half. It's a meal. Also good for an after-school snack.

Serves one.

### *Dilly Brussels Sprouts*

3 cups Brussels sprouts
½ cup water
1 cup yogurt
1 tablespoon minced dill tops

Cook sprouts in boiling water eight to ten minutes. Drain and save water, if any, for soup. Stir in yogurt. Top with dill.

Serves four to six.

### *A Neat Trick with Leftover Brussels Sprouts*

Drop the sprouts into opened jars of dilled beans or dilled cucumber pickles. The pickling liquid and dill flavor take a week to seep into the sprouts. They are delicious.

### *Brussels Sprouts Parmesan*

3 cups Brussels sprouts, cooked 8 minutes in ½ cup water
3 tablespoons butter
½ cup freshly grated Parmesan cheese
¼ cup toasted bread crumbs

Place cooked sprouts in baking dish and dot with butter. Top with grated cheese and bread crumbs. Bake at 350 degrees F for ten minutes.

Serves four to six.

### *Orange-Glazed Carrots*

¾ cup orange juice
1 tablespoon arrowroot flour
1 tablespoon honey
8 medium-size carrots, cooked
1 teaspoon grated orange rind
1 tablespoon butter

Slowly stir orange juice into arrowroot flour. Add honey and cook over low heat until thick, stirring to keep it from sticking. Pour sauce over cooked carrots and heat gently for five minutes so flavors blend. Add orange rind and butter last.

Serves four to six.

The secret of serving sweet, tasty carrots is to use as little cooking water as possible. When the carrots are tender, the water should have just evaporated. Better to watch them and add more hot water than to drown out their flavor.

### *Carrot-Celery Casserole*

6 large carrots, sliced
6 stalks celery, sliced
¼ cup water
1½ cups cream sauce (see page 281)
1 cup cubed stale bread or toast

Cook carrots and celery together in water until almost tender. Use any water left in the pan when making the cream sauce. Stir vegetables into cream sauce and put into a baking dish. Scatter bread cubes on top. Bake at 350 degrees F for 30 minutes.

Serves four to six.

### *Celery Cheese Casserole*

2 cups celery cut in ½-inch pieces
water
4 cups bread cubes
3 cups grated sharp cheese
3 eggs, slightly beaten
2 cups milk
pinch of thyme
½ teaspoon powdered mustard
pan of hot water

Cook celery in small amount of water until nearly tender. Butter a casserole dish and layer the celery, bread, and cheese in the pan: first a layer of bread cubes, then a handful of cheese, then a layer of celery. Next more bread cubes and cheese, then the rest of the celery, and top it off with more bread cubes and cheese. Pour mixture of eggs, milk, and herbs over the top and let the casserole stand for an hour before baking it. Then bake at 350 degrees F in a pan of hot water for about 75 minutes.

Serves six to eight as a side dish, four as a main dish.

### *Baked Cauliflower*

1 head cauliflower
1/4 cup melted butter
freshly ground pepper
2 tomatoes, quartered and cored
1 cup grated sharp Cheddar cheese

Break cauliflower head into pieces and toss in melted butter. Season with pepper. Arrange in baking pan with tomatoes. Bake at 350 degrees F for 25 minutes. Sprinkle cheese over top and return to oven for 5 more minutes.

Serves four to six.

### *Cheesy Baked Celeriac*

1 celeriac root
water for steaming
3/4 cup grated sharp cheese
1 1/2 cups cream sauce (see page 281)
2 slices of bread, cubed

Steam celeriac root for 30 minutes. Peel and cut into slices. Add cheese to hot cream sauce and stir into sliced celeriac. Top with bread cubes. Bake at 350 degrees F 45–60 minutes.

Serves four to six.

### *Leeks in Lemon Butter*

2 leeks
water for steaming
3 tablespoons butter
3 tablespoons lemon juice

Steam leeks until tender, about 15 minutes. Drain, and save water for soup. Melt butter. Add leeks to pan with butter and pour lemon juice over. Heat briefly and serve.

Serves four.

### *Baked Eggplant*

1 eggplant, cut crosswise into ½-inch slices
⅓ cup salad oil
4 tablespoons lemon juice
½ teaspoon oregano
⅔ cup grated sharp cheese

Marinate eggplant slices in oil/lemon juice/oregano mixture for 20 minutes. Then spread slices in shallow baking dish and bake at 350 degrees F for 30–40 minutes, adding cheese for the last 5 minutes.

Serves four.

### *Stuffed Eggplant*

1 eggplant, cut in half lengthwise
1 small onion, chopped
1 small bell pepper, chopped
2 tablespoons salad oil
1 stalk celery, diced
4 fresh tomatoes or 1 pint canned tomatoes
2 slices bread, cubed
½ cup grated sharp cheese

Scoop out and save eggplant pulp, leaving a ¼-inch layer of pulp on the skin. Sauté onion and pepper in oil until limp. Add celery, tomatoes, and chopped eggplant pulp. Cover and cook until tender, about 10 minutes. Mix in cubed bread. Pack vegetable mixture into the two eggplant shells. Top with cheese and bake at 350 degrees F 20–30 minutes.

Serves six.

### *Horseradish*

Sliced horseradish isn't very strong smelling. Grating releases its pungency. To make about a pint of grated horseradish relish, dig four carrot-size roots. Scrub but don't peel them. Chop the roots in inch-long segments. Drop them into a blender ⅓ cup at a time. Pour ¼ cup vinegar into the blender. Blend roots until all are finely chopped. Sometimes you must fish out the solid chunks, remove already grated material, add more vinegar and the unbroken chunks, and blend again. Hold your breath when scooping the grated horseradish out of the blender! The elderly man who grates horseradish at our local farmers' market keeps a fan going to blow the fumes away from him. Not a bad idea.

### *Onions, Peppers, and Bean Sprouts*

2 medium onions, chopped
2 bell peppers, green or red, chopped
3 tablespoons oil
2 cups bean sprouts

Sauté onions and peppers in oil for about eight minutes, until just tender. Add bean sprouts and stir just until sprouts begin to look transparent.

Serves four to six.

### *Onion Cheese Pie*

6 slices nitrate-free bacon
2 cups sliced onions
pastry to line bottom of 9-inch pie pan
soft butter
½ pound sharp cheese, grated
2 large or 3 medium eggs, slightly beaten
¾ cup milk
freshly ground black pepper

Fry bacon until crisp. Drain, and crumble into small pieces. Then sauté onions in bacon fat until tender. Butter piecrust lightly with soft butter and spread onions over the bottom. Then sprinkle grated cheese over onions and put bacon bits on top. Mix eggs and milk and pour over onion/cheese/bacon layers. Sprinkle with pepper. Bake at 400 degrees F for 10 minutes. Then reduce oven heat to 325 degrees and bake for 25–30 minutes. Serve hot.

Serves four to six.

### *Buttered Parsnips*

3 hefty parsnips
2 tablespoons water
3 tablespoons butter

Scrub parsnips and cut out joints where dirt lodges, but do not peel the roots. Slice crosswise if roots are tender; lengthwise, discarding hardest part of core, if roots are tough. Steam 1–2 minutes in boiling water. When water is evaporated, turn parsnips out into skillet and add butter. Cover and cook over low heat for 15–20 minutes, until tender, and slightly browned. Parsnips are especially good with lamb and peas.

Serves four to six.

### *Baked Parsnips*

3 cups sliced parsnips
2 tablespoons water or orange juice

Scrub and slice parsnips without peeling them. Set in baking dish and sprinkle water or juice over them. Cover, and bake at 350 degrees F for 20–30 minutes.

Serves four.

### *Mike's Potato Pancakes*

6 fist-size potatoes, peeled and grated
1–2 tablespoons flour
1 small onion, grated
1 egg, slightly beaten
oil for frying

Grated potatoes will turn brown but that doesn't matter. Pour off some of the extra liquid that accumulates after grating. Stir in flour, onion, and beaten egg. Use just enough flour to help to hold mixture together. Spread cakes out in hot oil in frying pan. Brown on both sides. Serve with yogurt and applesauce. Dill pickles are a good accompaniment, too.

Serves four.

### *Quick Shredded Turnips*

4 medium-size turnips, chilled
⅓ cup cooking oil

Shred turnips on four-sided kitchen grater. Brown in hot oil, turning after 5 minutes. They cook quickly. Allow 10–12 minutes.

Serves four.

### *Winter Radish Relish*

1 winter radish — Black Spanish, China Rose, or other
1 medium-size beet, cooked
1 tablespoon lemon juice

Grate radish and beet. Mix together and add lemon juice. Chill.

Makes about 1½ cups.

### *Harvester's Rutabaga Casserole*

1 medium large rutabaga, peeled and chopped
2 cooking apples, peeled and sliced
pinch of cinnamon
pinch of nutmeg
2 tablespoons honey
1 tablespoon butter (optional)

Steam cut-up rutabaga in covered pan for 10–12 minutes. Alternate layers of rutabaga and sliced apples in casserole dish. Sprinkle spices over the top, spoon on honey, and dot with butter. Cover dish and bake at 350 degrees F for 30–40 minutes.

Serves five to six.

### *Rutabaga Pudding*

3 cups mashed, cooked rutabagas
½ cup milk
2 eggs, slightly beaten
1 tablespoon honey
½ teaspoon nutmeg
1½ slices buttered bread, cubed

Mix rutabaga with milk, eggs, and honey. Sprinkle nutmeg on surface. Spread bread cubes on top. Bake at 350 degrees F for one hour.

Serves four to six.

### *Salsify Patties*

1 slice bread
1/4 cup milk
6 salsify roots, cooked and roughly mashed with fork
1 egg
oil (optional)

Soak bread in milk. Mix mashed salsify, egg, and milk-soaked bread. Shape into patties. Brown on both sides in hot oil or bake in 350 degree F oven in buttered shallow pan for 30 minutes. Serve with beets and green salad.

Serves four.

### *Classic Baked Squash*

Steam whole squash in covered kettle for 20–30 minutes. This makes it easier to cut. Whack the squash in half the long way. Rake out the seeds and stringy pulp. Place squash cut side down on greased baking sheet and bake at 350 degrees F for an hour, or until tender. Slice pieces crosswise to serve, unless squash is small.

### *Fruited Squash Rounds*

1 squash neck
1 tablespoon frozen orange juice concentrate
1 cup unsweetened crushed pineapple, drained
2 tablespoons butter

Cut rounded end from squash and save for another meal or for making pie. Steam neck end of squash for 20 minutes. Cut neck into 1/2-inch-thick slices. Put slices in greased baking pan. Mix orange juice concentrate with crushed pineapple and spread over the slices. Dot with butter. Bake at 350 degrees F for 50–60 minutes.

Serves six.

### *Squash with Creamy Greens*

- 2 seed-cavity ends of butternut squash, 2 acorn squash, or 2 other varieties of hollow squash
- 1 onion
- 1 tablespoon oil
- 2 cups cooked greens, such as spinach, beet tops, wild greens
- 1 cup cream sauce (see page 281)
- pecan meats, chopped

Steam, cut, and bake squash as in Classic Baked Squash above, but bake for only 40 minutes. Sauté onion in oil. Remove squash from oven, spoon greens mixed with cream sauce and onion into squash cavities, and return to oven for 20 minutes. Top with chopped pecans.

Serves six to eight.

### *Fried Green Tomatoes*

- 4 large mature green tomatoes
- ¼ cup whole wheat flour
- 1 egg, slightly beaten
- 2 tablespoons milk
- ¼ cup wheat germ
- cooking oil

Slice tomatoes, discarding stem and blossom ends. Dredge slices in flour. Mix egg and milk. Dip tomato slices in egg and milk mixture. Finally, sprinkle tomato slices with wheat germ on both sides. Fry on both sides in hot oil. Some people would be disappointed if there were not enough green tomatoes left on the vine to have this dish at least once a season.

Serves four to six.

### *Scalloped Turnips*

4 medium-size turnips, sliced very thin
½ cup hot milk
½ cup toasted sunflower seeds

Spread thin-sliced turnips in shallow baking dish. Pour hot milk over them. Bake at 400 degrees F for 40–60 minutes, until tender. Sprinkle sunflower seeds on top just before serving.

Serves four.

### *Sweet Potato Pudding*

2 tablespoons honey
⅓ cup orange juice
1 teaspoon grated orange rind
1½ cups cooked, mashed sweet potatoes
2 large or 3 small eggs

Stir honey, juice, and rind into potatoes. Separate eggs. Stir yolks into potato mixture. Beat whites until stiff and fold into the sweet potato/juice/honey mixture. Bake at 350 degrees F for 40 minutes.

Serves four.

# Unprocessed Crock Pickles

### *Sauerkraut*

Homemade sauerkraut is much less sharp, and at the same time far more flavorful, than any canned or bagged kraut you can buy. Home-crocked kraut has other advantages too. It lacks the sodium benzoate that's been added to the plastic-bagged kraut you find among the meats at the store. And it can be eaten uncooked. Canned sauerkraut has, of course, been heated in processing. You don't know what a Reuben sandwich *is* until you drape it with crisp shreds of fresh homemade kraut!

For years and years before canning and freezing were possible or even heard of, gardeners have used the last of the solid, frost-silvered cabbages to make sauerkraut. Salt and the lactic acid that develops as the cabbage ferments act as preservatives. Making kraut was, at first, a matter of economy and necessity.

As it turned out, the kraut tastes good. Its tang makes bland meals taste better. The lactic acid generated in the fermentation process is good for the body, too — one of our favorite examples of folk wisdom. Lactic acid, like yogurt, buttermilk, and acid fruits, helps to dissolve the iron in iron-rich foods so that it can enter the bloodstream.

To make a batch of sauerkraut, you'll need the following equipment:

- Cabbage shredder or sharp knife.
- Stomper — any clean wooden mallet or plunger — or a clean bottle or two-by-four.
- Scales — or a good guesstimator.
- Measuring spoons.
- Glazed crocks or gallon glass jars or clean barrel. Avoid metal containers. The acid in the kraut reacts with them.
- Large clean surface on which to shred the kraut.

Ingredients are simple: cabbages and salt. Discard the blemished outer leaves of the cabbage. Save the remaining large loose inner leaves for covering the kraut. If the cabbages are large, cut them in half lengthwise. Adjust your slaw cutter for a fine slice, and rub the cabbage across the blade. Slaw that is finely shredded will cure more thoroughly.

As the shredded cabbage piles up in mounds on the table, your helper can be weighing and salting it, leaving you free to saw away at the slaw board. Weigh a clean bowl or adjust the scale to compensate for its weight. Pile the shredded cabbage in the bowl and weigh it. Then dump it into the crock.

We salt the shredded cabbage at the rate of one tablespoon of salt for every two pounds of cabbage. This is less salt than many recipes call for, but the kraut is delicious and it keeps very well.

Stomp the salted cabbage shreds with your plunger, starting when the crock is about one-third full. It will seem slow at first but soon the juice will start to rise. By bruising the mix with your stomper you hasten the interaction of salt and cabbage. Once the juice starts to flow it rises steadily. More shredded cabbage may be added after the juice starts to flow. Leave a two-to-three-inch head space at the top of the crock for further rising of the juices during fermentation.

When the crock is full, cover it with the clean cabbage leaves you've reserved, invert a plate over the juicy cabbage, and weigh the plate down with a clean heavy object to keep the cabbage submerged in the liquid. We use a half-gallon jar filled with water for a weight. The traditional weight is a well-scrubbed rock.

Curing temperature determines the quality of the kraut. Researchers report that the best temperature for initial fermentation is 59 to 68 degrees F maintained for two weeks, with a gradual drop to around 32 degrees after four weeks. When we make kraut in the fall it usually cures at around 60 to 75 degrees and we are entirely satisfied with the result. Above 86 degrees, though, the kraut tends to ferment too rapidly, producing a soft and often sour product. Below 50 degrees, fermentation slows considerably. It stops at 32 degrees.

You can tell that your kraut is fermenting if bubbles are rising in the liquid. Sometimes the liquid overflows. We set our crocks on pads of newspaper in shallow pans to catch the overflow. When bubbling stops, the kraut has finished fermenting. Then we fork off the large leaves and skim off the mold from the top surface of the kraut. We replace the plate and weight with clean ones and keep the kraut in a cold place — 40 degrees F or below. Kraut will keep for a month or so in a warmer place — 45 to 50 degrees — but at those temperatures spoilage organisms become more active.

Cover the stored sauerkraut with a clean cloth or plastic sheet to keep out dust. And enjoy the privilege of dipping into the crisp tangy brew as sausage sizzles on the stove.

Turnips, rutabagas, and snap beans may also be preserved in brine by the same procedure. Shred the root vegetables, salt and stomp them just as you would the cabbage. Beans should be sliced the long way —

French-style. Beans don't always produce as much juice as the cabbage family vegetables. If you need more brine to keep the beans covered, dissolve one part of salt in eight parts of water, by weight, and add to the crock.

### *The Pickle Crock*

Making pickles is a great way to preserve that last burst of energy from the pepper plants and tomato and cucumber vines. When fermented in brine, these vegetables develop a uniquely delicious flavor.

You'll need the following:

- 1 heaping peck of washed green vegetables: mature green tomatoes, cucumbers or peppers. Do not mix different vegetables in the same crock unless individual pieces are the same size; small portions cure more quickly than large ones. We quarter cucumbers and green tomatoes. Some people leave them whole.
- 8–10 dill umbels, flowering or gone to seed
- 1½ cups uniodized salt
- 6 cups vinegar
- 12 cups water (3 quarts)
- 10–15 clean grape leaves (these make the pickles crisp)

Line a clean crock or glass jar with half the dill heads and grape leaves. Pack in the vegetables up to the halfway mark. Dissolve the salt in the water and stir in the vinegar. Pour this brine over the vegetables. Add another layer of dill and grape leaves and any other spices you wish to include. Then put in the rest of the tomatoes, cukes, or whatever, pour brine over them, top with grape leaves, and put a clean plate and weight over the crock to keep the vegetables submerged, as directed on page 271. You can put a disc of wood on top of the crock instead of a plate but don't use pine; it imparts a resinous flavor to the pickles.

When possible, pickle cucumbers within a day after harvesting. When kept longer than that, they tend to lose their small store of natural sugar which is needed for fermentation by lactic bacteria.

Other good flavorings to add include:

- Peeled garlic cloves
- Hot peppers

- Horseradish leaves
- Tarragon
- Parsley
- Horseradish root
- Celery
- Black currant leaves
- Mustard seeds
- Tiny whole onions

Keep the pickles in a cool place — ideally 68 to 72 degrees F — during the period of fermentation, which takes from two to four weeks. If you've cut your vegetables in small pieces, check them at the end of two weeks to see whether they're ready. Skim the crock before you remove some pickles, being careful not to mix the mold in with the liquid. Don't worry about the natural whitish scum that remains on top of the surface. Just rinse off the pickles. Once fermentation is complete, the pickles should be kept as cold as possible — at 40 degrees F or lower. Cover them lightly to keep out dust.

## Baked Goods

### *Hickory Nut Granola*

1 cup honey
1 cup cooking oil
8 cups coarse rolled oats
2 cups wheat germ
1 cup bran
2 cups sunflower seeds
1½ cups hickory nuts, reserved

Heat honey and oil together until honey is liquid. Combine remaining ingredients, reserving nuts. Pour honey mixture over grain mixture, combining thoroughly. Bake in greased pan at 325 degrees F for 30–40 minutes. Stir several times during baking period. Remove from oven and mix in hickory nuts.

Makes about 11 cups.

### *Pear Bread*

2 pears
½ cup butter
⅓ cup honey
2 eggs
2¼ cups flour
2 teaspoons baking powder
¼ teaspoon baking soda
¼ cup sour milk or yogurt
1 teaspoon grated lemon rind
1 cup granola
½ cup raisins
3 tablespoons granola

Grate 1½ pears. Save remaining half for topping. Soften butter and add honey and eggs. Then add grated pears. Next stir in dry ingredients alternately with sour milk. Stir in lemon rind. Add granola and raisins last. Spoon batter into greased, floured 9-by-5-by-3-inch pan. Slice reserved pear half in thin wedges and poke into batter. Sprinkle 3 tablespoons of granola over the top of the bread. Bake at 350 degrees F for one hour. Slices best the day after baking.

Makes one large loaf.

### *Orange Bread*

1 whole orange, cut into eighths, and seeded
½ cup water
⅔ cup raisins
2 eggs
⅓ cup honey
2½ cups whole wheat flour
2½ teaspoons baking powder
nut meats (optional)

Whirl orange sections, one at a time, with water in blender until coarsely chopped. Add raisins and buzz very briefly, just enough to break raisins up a bit. Then add eggs and honey to blender and blend. Mix liquified ingredients in blender with flour and baking powder. Turn into greased, floured bread pan or square cake pan. Top with pecans or walnuts if desired. Bake at 350 degrees F for about an hour for loaf pan, 35-40 minutes for shallow pan.

Makes one loaf.

### *Sweet Potato Bread*

- 2 packages active dry yeast
- 1/4 cup warm water
- 1 cup mashed, cooked sweet potatoes
- 2 eggs
- 1/4 cup vegetable oil
- 1 tablespoon honey
- 2 3/4 cups scalded, cooled milk (or water)
- 6-6 1/2 cups flour
- 1 teaspoon diastatic malt (optional, but gives loaf excellent texture)
- 1 egg, beaten
- 1 tablespoon water

Soften yeast in warm water. Stir yeast into sweet potatoes. Then add eggs, oil, and honey. Next stir in milk or water. Gradually mix in flour, and malt if used, reserving last 2 cups of flour for kneading dough. Knead until dough surface loses its stickiness. Put dough in greased bowl, cover, and let rise in a warm place for an hour. Then poke fist into dough to deflate, divide in half, and shape each half into a loaf. Put into greased, floured bread pans (or muffin tins for rolls) and let rise until not quite doubled (30-50 minutes depending on temperature of kitchen). Brush tops of loaves with egg beaten with 1 tablespoon of water. Bake at 350 degrees F 45-55 minutes.

Makes two loaves or three to four dozen rolls.

### *Nancy's Black Walnut Waffles*

- 1 cup whole wheat flour or mixed-grain flour
- ½ cup buckwheat flour
- ½ cup soy flour
- 4 teaspoons baking powder
- ⅓ cup vegetable oil
- 3 eggs, separated
- 1½ cups milk
- 1-1½ cups black walnuts

Mix flours and baking powder. Stir in oil, egg yolks, and milk. Then add nuts. Finally, beat egg whites until stiff and fold them into waffle mixture. Bake in waffle iron. Serve with real maple syrup.

Makes about sixteen waffles.

# Desserts

### *Carrot Steamed Pudding*

2 eggs
1 cup grated raw carrot
1 cup grated raw potato
1 cup finely chopped peeled cooking apple
1/4 cup honey
1/3 cup molasses
2 cups whole wheat flour
1 1/2 teaspoons baking powder
3/4 teaspoon baking soda
1 teaspoon cinnamon
1 teaspoon nutmeg
1/2 teaspoon cloves
1 cup raisins
3/4 cup dates, chopped

Mix eggs, grated carrot, potato, and apple with honey and molasses. Stir in flour, baking powder, baking soda, and spices. Add raisins and dates last. Spoon into greased, floured pudding molds or cans and cover with foil. Steam on rack in one inch boiling water in covered kettle for 1 1/2 hours if divided into two or more cans, 2 1/2 hours if all in one container. Serve with your favorite sauce. Here's ours:

### *Cream Cheese Sauce*

1 8-ounce package cream cheese, or homemade cheese made from yogurt dripped through cheesecloth overnight
1/4 cup honey
1 teaspoon vanilla or 1 tablespoon frozen orange juice concentrate

Mash cream cheese and work honey into it. Add vanilla or orange juice concentrate.

***Pumpkin Custard***

- ½ teaspoon nutmeg
- ½ teaspoon allspice
- 1 teaspoon cinnamon
- 1 tablespoon flour or cornstarch
- 1½ cups mashed cooked pumpkin (or squash)
- ½ cup honey
- 3 eggs, slightly beaten
- 1½ cups milk

Stir spices into flour and mix with pumpkin. Then add honey, beating till smooth. Combine eggs and milk, and slowly stir into pumpkin mixture. Ladle into custard cups. Set cups in pans of water and bake at 350 degrees F for about one hour.

Makes about three cups.

***Pumpkin Pie*** (for which I almost always use squash)

Mix together all ingredients listed above for custard and pour into a 10-inch pie pan lined with piecrust. Bake at 450 degrees F for 10 minutes. Then reduce heat to 350 degrees and bake for an additional 50 minutes, or until edges are firm. Center may still be soft but will cook on standing. My neighbor Gwen Brandt gave me this recipe, and I have abandoned all other pumpkin pie recipes in favor of Gwen's.

Makes one ten-inch pie.

***Dip for Melon Squash Fingers***

- ½ pint yogurt
- ¼ cup chopped crystallized ginger

Mix and chill. Serve with slices of melon squash. Add side dishes of toasted sunflower and pumpkin seeds.

### *Pear Pie*

½ teaspoon nutmeg
dash of cinnamon
4 tablespoons flour
6 cups peeled, sliced ripe pears
½ cup honey
1 tablespoon lemon juice
pie shell and top pastry

Mix spices with flour and fork flour mixture gently into pears. Drizzle honey over all. Sprinkle with lemon juice. Turn into pastry-lined pie pan and top with pastry. Cut vents in top pastry. Bake at 425 degrees F for 15 minutes. Then reduce heat to 350 degrees and bake another 35–45 minutes. Good served warm.

Serves six.

### *Quince Applesauce*

Cook one or two quinces with each peck of apples you cook for sauce. Adds flavor. Sweeten as usual. We add honey to taste after cooking.

### *Apple Crisp*

7 or 8 cooking apples, peeled
½ cup butter
¾ cup date sugar (dried, ground-up dates — available in natural foods stores)
¾ cup oatmeal
½ cup whole wheat flour
1 teaspoon cinnamon

Slice peeled apples into buttered baking dish. Melt butter and stir in date sugar, oatmeal, flour, and cinnamon. Mix with fork until crumbly. Spread crumb mixture over apples in pan. Bake at 350 degrees F for 35–40 minutes. Makes the house smell heavenly for the after-school crowd.

Serves four to six.

# Basic Recipes

Here are recipes for foods or components suggested several times in the recipe section. We should add here that we use whole grain flour for all of the recipes in this section — whole wheat for yeast breads and, for most other food, a twelve-grain flour we buy from Walnut Acres.

### *Mayonnaise*

2 egg yolks
½ teaspoon dry mustard
1 tablespoon honey
2 tablespoons vinegar
1 tablespoon lemon juice
1 cup salad oil

Blend all ingredients except oil in blender until well combined. Then add oil very, very slowly, while running blender constantly. Keep ingredients as cold as possible.

Makes about 1½ cups.

### *Cooked Mayonnaise* (Pennsylvania Dutch Style)

½ cup milk
1 teaspoon cornstarch or arrowroot flour
1 egg
¼ cup vinegar
⅓ cup honey

Stir milk slowly into cornstarch or arrowroot. Beat egg and mix all ingredients together in saucepan. Cook over low heat, stirring steadily, until it thickens.

Adapted from my Great-Aunt Carrie's awfully sugary recipe.

Makes about ⅔ pint.

### *My Favorite Piecrust*

- 1½ cups natural lard (not hydrogenated) or 1¾ cups butter or margarine
- 4 cups flour
- 1 egg
- ½ cup hot water
- 1 teaspoon honey
- 1 tablespoon vinegar

Cut shortening into flour until lumps are pea size. Then fork egg and liquids together and add them gradually to the flour/shortening mixture. Chill before using. Freezes well.

Makes enough for five single pie shells or two doubles and one single.

### *Cream Sauce*

- 2 tablespoons butter
- 2 tablespoons oil (or use all butter and no oil)
- ¼ cup flour
- 2 cups milk

Melt butter over low heat. Add oil if desired. Remove pan from heat and blend in flour. Gradually stir milk into flour/butter mixture; if you add milk all at once the sauce will be lumpy. Cook, stirring, until thickened.

Makes about two cups.

# Bibliography


Abraham, George. *The Green Thumb Book of Fruit and Vegetable Gardening.* Englewood Cliffs, N.J.: Prentice-Hall, 1970.

Alth, Max. *Working with Concrete and Masonry.* New York: Harper and Row, 1978.

Beck, Charles F. "How to Store Your Food Crops, Part 2, Building the Storage Room." *Organic Gardening and Farming,* September 1971, pp. 84-93.

Beeson, Kenneth C. "Soil Management and Crop Quality." In *Soil,* the *USDA Yearbook,* 1957, pp. 258-67.

Biancardi, Tony. "Storing Irish Potatoes." *Organic Gardening,* September 1978, pp. 54-55.

Boswell, Victor. "Vegetables." In *Soil,* the *USDA Yearbook,* .1957, pp. 692-98.

Boynton, Damon, and Magness, John R. "Soil Management for Orchards." In *Soil,* the *USDA Yearbook,* 1957, pp. 699-710.

Bubel, Nancy. "A Consumer's Guide to Winter Squash Varieties." *Organic Gardening and Farming,* January 1978, pp. 69-73.

———. *Country Journal's Complete Handbook of Home Milk Receipts.* Manchester, N.H.: Country Journal Publishing, 1977, pp. 28-29.

———. *The Seed-Starter's Handbook.* Emmaus, Pa.: Rodale Press, 1978.

———. "Winter-Keeping Vegetables." *Country Journal,* September 1978, pp. 37-39.

Bush, Raymond. *Harvesting and Storing Garden Fruit.* London: Faber and Faber, 1947.

Carleton, R. Milton. *The Small Garden Book.* New York: MacMillan, 1971.

———. *Vegetables for Today's Gardens.* Princeton: D. Van Nostrand, 1967.

Cox, Jeff. "Potash: The Plant-Growth Catalyst." *Organic Gardening and Farming,* June 1976, pp. 138-41.

Dickey, Esther. *Passport to Survival.* Salt Lake City: Bookcraft, 1969.

Ellison, Susan. "Store and Enjoy Your Garden's Produce All Winter." *Natural Gardening,* October 1972, pp. 26-29.

*Farm and Household Cyclopedia.* New York: F. M. Lupton, 1885.

Forsyth, F. R.; Lockhart, C. L.; and Eaves, C. A. "Home Storage Room for Fruits and Vegetables." Ottawa: Canada Department of Agriculture, 1978.

Garden Way Bulletin, "Building and Using Your Root Cellar." Charlotte, Vt.: Garden Way, 1978.

Gaus, Arthur; DiCarlo, Henry; and Zurowest, Rudy. "Vegetable Harvest and Storage Fact Sheet." USDA, n.d.

Gough, Robert E. "Quince Culture." *Country Journal,* October 1978, pp. 106-7.

Haard, Norman F., and Salunkhe, D. K., ed. *Symposium: Postharvest Biology and Handling of Fruits and Vegetables.* Westport, Conn.: Avi Publishing Co., 1975.

Huff, Darrell. *How to Work with Concrete and Masonry.* New York: Harper and Row, 1976.

Isenberg, F. M. R. "Storage of Home Grown Vegetables." Mimeographed. Ithaca, N.Y.: Cornell University, 1973.

Ivins, Lester S. *Garden Crops: Production and Preservation.* New York: Rand McNally, 1919.

Jennings, Joan. "Foods That Fight Inner Pollution." *Prevention,* May 1975, pp. 56-61.

Kellogg, Charles E. "Home Gardens and Lawns." In *Soil,* the *USDA Yearbook,* 1957, pp. 665-88.

Kronschnabel, Darlene. "Root Cellars Are Great Harvest-Holders." *Organic Gardening and Farming,* September 1973, pp. 60-62.

Lappé, Frances Moore. *Diet for a Small Planet.* New York: Friends of the Earth/Ballantine, 1971.

Lear, Leonard. "The 'Honest Carbohydrates' — Great for Diabetics, Good for Everyone." *Prevention,* August 1976, pp. 157-66.

Logsdon, Gene. *Small-Scale Grain Raising.* Emmaus, Pa.: Rodale Press, 1977.

Minnich, Jerry. "Energy-Free Food Storage." *Countryside,* October 1977, pp. 15-22.

*The Moon Sign Book.* St. Paul, Minn.: Llewellyn Publications, 1975.

*The Mother Earth News* Staff. "Can You Really Store Fresh Eggs for a Year or More Without Refrigeration?" *The Mother Earth News,* November/December 1977, pp. 170-71.

Mueller, Jo. *Growing Your Own Mushrooms.* Charlotte, Vt.: Garden Way, 1976.

*Organic Gardening and Farming* Staff."Eat Raw Foods." *Organic Gardening and Farming,* August 1977, p. 129.

———. "Eating to Prevent Cancer." *Organic Gardening and Farming,* May 1977, p. 126.

*Prevention* Staff. "Treat Your Body to More Raw Food." *Prevention,* July 1977, pp. 111-16.

Robinson, Ed and Carolyn. *The Have-more Plan.* New York: MacMillan, 1947.

Rodale, J. I., and Staff. *The Encyclopedia of Organic Gardening.* Emmaus, Pa.: Rodale Press, 1959.

———. *How to Grow Vegetables and Fruits by the Organic Method.* Emmaus, Pa.: Rodale Press, 1961.

Salunkhe, D. K., ed. *Storage, Processing and Nutritional Quality of Fruits and Vegetables.* Cleveland: CRC Press, 1974.

Shanly, Eupha. "The Farm Cellar." *Poor Joe's Pennsylvania Almanack,* 1979, p. 86.

Shirakov, E. P. *Practical Course in Storage and Processing of Fruits and Vegetables.* Translated by L. Markin. First published in Moscow, 1964. Republished for USDA and National Science Foundation by the Israel Program for Scientific Translations, 1968.

Sloane, Eric. *Diary of an Early American Boy.* New York: Funk and Wagnalls, 1962.

Smith, Edwin. *Storage of Fruits and Vegetables.* Chicago: American Institute of Agriculture, 1923.

Swain, Roger. "Rotten Apples." *Horticulture,* November 1978, pp. 12-16.

———. "Storing Summer Surplus." *Horticulture,* September 1978, pp. 37-41.

Thompson, Walter A. "Storing Fruits and Vegetables." *Farmstead,* n.d., p. 56.

Thurber, Nancy, and Mead, Gretchen. *Keeping the Harvest.* Charlotte, Vt.: Garden Way, 1976.

Tiedjens, Victor. *The Vegetable Encyclopedia.* New York: Avenel Books, 1943.

U.S. Department of Agriculture. "Storing Vegetables and Fruits in Basements, Cellars and Pits." Home and Garden Bulletin No. 119. Washington, D.C.: Government Printing Office, 1970.

Wallner, Stephen, and Ferretti, Peter A. "Post Harvest Handling and Storage of Vegetables and Berries for Fresh Market." University Park, Pa.: The Pennsylvania State University, 1978.

Weaver, John E., and Brewer, William E. *Root Development of Vegetable Crops.* New York: McGraw-Hill, 1927.

Wigginton, Eliot, ed. *The Foxfire Book.* New York: Doubleday, 1972.

Wilcox, Louis V., Jr. "The Truth About Potatoes." *Farmstead,* Winter 1978, pp. 18-24.

Wolf, Ray, ed. *Managing Your Personal Food Supply.* pp. 219-66. Emmaus, Pa.: Rodale Press, 1977.

# Sources

**Nut Cracker for Black Walnuts and Hickories:**

Potter Walnut Cracker Co.
Box 930
Sapulpa, OK 74066

**Season-Extending Mini-Greenhouse:**

Guard 'n Gro
St. James, NY 11780
Brochure: 25¢

**Oyster and Shiitake Mushroom Logs:**

Thompson and Morgan
Box 100
Farmingdale, NJ 07727

**Fruits and Nut Trees:**

Vernon Barnes and Son Nursery
P.O. Box 250-L
McMinnville, TN 37110

Bountiful Ridge Nurseries Inc.
Princess Anne, MD 21853

Henry Leuthardt
East Moriches, NY 11940

Mellingers' Inc.
North Lima, OH 44452

J. E. Miller Nurseries, Inc.
Canandaigua, NY 14424

Southmeadow Fruit Gardens
2363 Tilbury Place
Birmingham, MI 48009
(Specializes in old and rare varieties.)

Stark Brothers
Louisiana, MO 63353

Worley's Nurseries
Rt. 1
York Springs, PA 17372

**Twelve-Grain Flour:**

Walnut Acres
Penns Creek, PA 17862

**Vegetable Seeds:**

Burgess Seed and Plant Co.
P.O. Box 221
Galesburg, MI 49053

W. Atlee Burpee
Warminster, PA 18974

P.O. Box B-2001
Clinton, IA 52732

and

6350 Rutland Avenue
Riverside, CA 92501

Comstock, Ferre and Company
263 Main Street
Wethersfield, CT 06109

DeGiorgi Company, Inc.
1411 Third Street
Council Bluffs, IA 51501

Epicure Seeds
Avon, NY 14414

Farmer Seed and Nursery Company
818 NW Fourth Street
Faribault, MN 55021

Henry Field Seed and Nursery Company
407 Sycamore Street
Shenandoah, IA 51602

Grace's Gardens
22 Autumn Lane
Hackettstown, NJ 07841

Gurney Seed and Nursery Company
Second and Capital
Yankton, SD 57078

Joseph Harris Company, Inc.
Moreton Farm
Rochester, NY 14624

The Charles Hart Seed Company
304 Main Street
Wethersfield, CT 06109

Herbst Brothers Seedsmen, Inc.
1000 North Main Street
Brewster, NY 10509

J. L. Hudson, Seedsman
P.O. Box 1058
Redwood City, CA 94064
(Send first class stamp for vegetable listing.)

Johnny's Selected Seeds (Rob Johnston)
Albion, ME 04910

J. W. Jung Seed Company
Randolph, WI 53956

Earl May Seed and Nursery Company
North Elm
Shenandoah, IA 51603

Nichols Garden Nursery
1190 North Pacific Highway
Albany, OR 97321

L. L. Olds Seed Company
2901 Packers Avenue
P.O. Box 1069
Madison, WI 53701

George W. Park Seed Company, Inc.
P.O. Box 31
Greenwood, SC 29647

Seedway, Inc.
Hall, NY 14463

Shades of Green
16 Summer Street
Ipswich, MA 01938

R. H. Shumway, Seedsman
628 Cedar Street
Rockford, IL 61101

Stokes Seeds, Inc.
P.O. Box 548
Buffalo, NY 14240

Thompson and Morgan
Box 100
Farmingdale, NJ 07727

Otis Twilley Seed Company
Box 1817
Salisbury, MD 21801

# INDEX

*Page numbers in italics indicate illustrations.*

## C

## R



## S